P9-AQN-509

THE ALMIGHTY CHANCE

World Scientific Lecture Notes in Physics, Vol. 20

THE ALMIGHTY CHANCE

Ya. B. Zeldovich
A. A. Ruzmaikin
D. D. Sokoloff

World Scientific
Singapore • New Jersey • London • Hong Kong

Published by

World Scientific Publishing Co. Pte. Ltd.
P O Box 128, Farrer Road, Singapore 9128
USA office: 687 Hartwell Street, Teaneck, NJ 07666
UK office: 73 Lynton Mead, Totteridge, London N20 8DH

THE ALMIGHTY CHANCE

ISBN 9971-50-916-4
9971-50-917-2 (pbk)

Printed in Singapore by General Printing Services Pte. Ltd.

PREFACE

This book, devoted to one of the directions of modern physics, namely, the emergence of order from chaos, is among the last works of an outstanding Soviet physicist, Yakov Borissovich Zeldovich. His interest in these problems stems from his studies of the combustion theory, heat transfer and magnetic-field generation in moving media, and the structure of the universe — work that he carried out over a long period from the thirties up to the eighties. The immediate basis of this book is formed by his papers on the theory of random media, especially those published in collaboration with S.A. Molchanov and the other two authors of this book (Zeldovich *et al.*, 1985, 1987a, b). These studies apply rather complicated mathematics, although the basic results are of quite general and simple nature. These studies led to a simple and brief discussion of the abovementioned problems as well as transparent connections between various phenomena.

Yakov Borissovich has applied all his enormous energy to persuade us that this is the very approach that is needed. At the moment of his untimely death, the text was at the final stage of preparation: the plans for all the chapters were worked out and the draft of almost the entire text had been prepared. We worked at several sections of the manuscript simultaneously and revised it together to reach the final form. Quite naturally, with this style of work, the book was not written consecutively from the first to the last page. In his last weeks, Yakov Borissovich wrote Chapter 3 and during his final days he was finishing Chapter 2. As usual, he wrote on a simple schoolboy's notebook. The last pack of these notebooks was given to us a few days before his death. We consider this work as planned by him a tribute to his memory.

Working on the text, we tried to put together the idea of a simple and compact book as well as the more subtle ideas of Ya.B. Zeldovich. In the text as it is, only in Chapter 9 did we demonstrate some of the technical complexities that arise in the application of the theory to specific problems. We hope that we will be able to fulfill our initial plan and to devote more detailed discussions on special mathematical problems in our next book (which will be written together with S.A. Molchanov, who has made substantial contributions to the mathematical formalism of intermittency theory).

We did not attempt to finish a small fragment at the end of Chapter 2 where Zeldovich exposed the mathematical formalism required for the chapter on the structure of the universe. The beginning of this fragment was written by him three days before his death and we just give his text here:

> 8. The General Random Homogeneous Distribution of Particles
>
> In the preceding sections, we considered the random distribution of molecules that would arise when the Lord populates a vessel with molecules in a purely random manner.
>
> The Lord fully expresses his omnipotence by giving free play to chance: any individual molecule is positioned at any place with equal probability. To be more exact, the probability of putting a molecule in the elementary volume dv is dv/V, where V is the total volume of the vessel. The Almighty would not miss and the total probability for a molecule to be put into any part of the vessel is unity. $\int dv/V = V/V = 1$.
>
> It would be a mistake to assert that exactly the same statistical properties have a distribution that arose due to this or that physical process even if it preserved *homogeneity* of space, i.e., all spatial parts statistically have the same properties.
>
> the end for 30/XI. 1987,
> but not the end (sic)

Everything said or made in the last hours of life inevitably acquires a symbolic meaning even though it was an ordinary everyday event. The last notes of Zeldovich are remarkable for his appeal to the Almighty supplemented at the margin by the proposition of discussing in the text the concept of hypostasis. Nevertheless, we are not tempted to overestimate this fact. As far as we know, Yakov Borissovich was not a religious person and the best we can do is to refer to the French writer Marcel Proust (Zeldovich always felt an intrinsic connection with French literature whose translator into Russian was his mother, a member of the Union of Soviet Writers) who wrote that the highest praise to God is the unbelief of a scholar who is sure that the perfection of the World makes existence of Gods unnecessary.

We consider the very last phrase more symbolic, "the end for 30/XI.87 but not the end". It is surprisingly close to the well-known *Vortsetzung folgt* at the very end of the Wilhelm Meister dialogue by Goethe, also one of the last writings of that great German writer.

The text was translated from Russian by Anvar Shukurov, also a pupil of Ya.B. Zeldovich. His comments helped to improve this presentation and his role goes beyond a translator's job.

We wish to express our deep gratitude to S.A. Molchanov, a co-author of many original papers whose physical aspect is discussed in this book. Several examples taken from the fields of science, in which the authors are not specialists, have been proposed to us by V.V. Dvornichenko, V.L. Egorov, M.S. Markov, D.D. Sokoloff, Jr., and Yu.V. Chernov. Special thanks are due to M.I. Bujakaite for her help in the preparation of the manuscript. The consideration of the VAAP representative, S.A. Sergeeva, is also gratefully acknowledged.

A.A. Ruzmaikin
D.D. Sokoloff

CONTENTS

CHAPTER

ONE

INTRODUCTION

The problems of physics of random media are always a source of novel ideas. This field of science is associated in a versatile way with real complexity of random media (as an ensemble of enormous numbers of molecules that form macroscopic bodies, or a plentitude of disordered motions typical of turbulence, etc.). Following the well-known aphorism of David Hilbert, we may only wonder how physicists can deal with a random medium though the subject may be too complicated for them. The simplest fruitful approach consists in the derivation and solution of equations for macroscopic bodies or averaged quantities, e.g., temperature which is essentially the averaged kinetic energy of molecular motion. A problem once proposed by *l'Académie Française* which considered heat propagation in solids for given boundary conditions has become a basis of classical mathematical physics. J.B. Fourier solved this problem in 1807 and 1811 by separation of variables. This solution brought about such important concepts as the Fourier series and integral, eigenvalues, and eigenfunctions. Later, methods of solution of other equations of mathematical physics were formulated, in particular equations of elastic oscillations and the Maxwell and Schrödinger equations.

Solving the heat conduction equation by following the approach of Fourier, one does not need any knowledge of the kinetic theory of heat. Temperature or, say, the concentration of diffusing admixture, can be considered as a continuous and smooth function of position and time. And still later on, in the early twentieth century, the possibility of solving statistical problems by means of differential equations and regular functions was considered a great development. The statistical mechanics of L. Boltzmann and J.W. Gibbs has allowed us to determine properties of equilibrium states, e.g., molecular distribution in a given potential force field. A. Einstein established the connection between the mean mobility of particles in the force field and their diffusion driven by the concentration gradient. This removed a mystical veil from the phenomenon of osmotic pressure which is so important for biological cells and life itself. L. Boltzmann, M. Planck, A. Fokker, P. Langevin, D. Hilbert and, more recently, N.N. Bogolyubov have developed the kinetic theory.

In the heat conduction problem the very nature of solution makes the approach of Fourier favourable. Higher modes rapidly decay and thereby a basic distribution of temperature is quickly established, practically independently of the initial distribution. The same result can be obtained with the help of Green's function,

$$\mathcal{G}(t, \mathbf{x}, \mathbf{y}) = (2\pi\kappa t)^{-3/2} \exp\left[\frac{-(\mathbf{x}-\mathbf{y})^2}{2\kappa t}\right],$$

which is especially convenient in unbounded domains. Here $\mathbf{x}$ and $\mathbf{y}$ are two spatial points, and κ is thermometric conductivity. The convolution of the initial distribution with this function smooths out and supresses singularities.[a]

However, there is an effect that introduces greater complexity into the process of heat conduction or diffusion. This is the motions in a medium where diffusion occurs. In a typical case the motion is turbulent and can be determined only statistically. As it was first shown by J. Taylor (1921), in the case of evolution of large-scale averaged quantities, e.g., the mean temperature, it can again be described by classical equations of mathematical physics with effective transfer coefficients (the turbulent diffusivity). But in the case of enhanced temperature spots (this was initially noticed by Gibbs, who considered the evolution of a drop of ink in water, and substantiated by D. Townsend and D. Batchelor (1950)), turbulence

[a] This property is far from trivial. Note that in the theory of oscillations, an initial singular state (delta-function) preserves its spectrum throughout the evolution provided the medium is ideal (without dissipation). In the corresponding generic boundary-value problem, the modes cease to be phased and what arises is the unsmooth, fractal spatial distribution of density and pressure. The Green's function of the wave propagation equation is also singular.

can make sharper gradients without increasing the maximal temperature. In other words, turbulence feeds higher Fourier modes and irregularities of an embedded scalar field. The rate of change of the integrals of the type $\int T^2 d^3\mathbf{x}$, that determines the rate of establishment of equilibrium distribution, is proportional to $-\kappa\int(\nabla T)^2 d^3\mathbf{x}$ (Zeldovich, 1937; Obukhov, 1949). In problems with heat sources the stationary heat flux is also expressed through the latter integral.

One more complication is revealed in the case of an embedded vector field. For instance, the evolution of the (vectorial) magnetic field embedded in a moving conducting fluid plays a decisive role in the problem of the origin of magnetism of the Earth, the Sun and many other cosmic objects. In the limit of high electric conductivity the field is frozen into the fluid, as H. Alfven and K.O. Kippenheuer put it. It moves together with the fluid particles (the field vector which connects two infinitesimally close particles varies proportionally with the distance between these particles). Owing to this specific property, a suitable motion of the medium does not only transfer the field, in contrast to the scalar case, but also provides continuous and prolonged growth of the magnetic field[b] which can be stopped only by back-action of the field on the motion, i.e., a nonlinear effect.

Nonlinear problems of heat conduction and vector field generation have a common characteristic feature: the formation of dissipative structures, i.e., compact regions of enhanced intensity. The relevant theory, pioneered by I. Prigogine and impressively popularized by H. Haken, was actively developed by many groups of scientists. Here are some fields where this theory has been succesfully applied: (a) biology — the formation of the structure of a living organism to propagate new species in the biosphere; (b) combustion — heat localization in nonlinear media, explosions, formation of blast waves and their (in)stability and structure; (c) evolution of the structure of the universe — formation of galaxies and their clusters from nearly uniform plasma whose existence at early stages of evolution can be considered as firmly established.

[b] Note that vorticity obeys the same equation as the magnetic field. However, an important distinction is that magnetic field can be considered embedded in an arbitrary velocity field while a given vorticity distribution completely determines the corresponding velocity field.

Let us discuss the latter problem in more detail. Fragmentation of nearly uniformly distributed matter into more or less isolated galaxies (conglomerates of stars) and galaxy clusters proceeds under the action of gravity. (The assertion that thermodynamic equilibrium corresponds to spatially uniform density and temperature — the "thermal death" — is true only in everyday or laboratory physico-chemical situations but not in the presence of long-range gravitational forces.) There are reasons to believe that in the initial distribution of matter short waves are suppressed by dissipative processes. Unstable growth of longer waves leads to the intersection of particle trajectories, i.e., to the formation of caustics similar to optical ones (Zeldovich, 1970; Shandarin and Zeldovich, 1984; Arnold, 1980). The result is the formation of surfaces where the density ρ is formally infinite. The mean squared density, $\langle \rho^2 \rangle$, is divergent; meanwhile, obviously, ρ can never be negative. Therefore, the distribution of deviations of the density from the mean value essentially differs from the Gaussian one: there are singularities, i.e., $\rho \to +\infty$, but nowhere is $\rho < 0$. This peculiarity is closely associated with the fact that in Fourier expansions of singularities (caustics) short-wavelength and high-frequency harmonics have large amplitudes, whereas in the initial spectrum the amplitudes of higher harmonics are zero or decay exponentially. The formation of singularities leads to a power-law decrease of the amplitudes with frequency. A detailed analysis of Shandarin and Zeldovich (1984) shows that the resulting structure has the form of a network from two-dimensional view and a cellular structure with thin dense walls in three-dimensional space.

However, the formation of intensity peaks or structures are possible even at the linear stages of evolution when the coefficients of the governing linear equation (5) are determined by properties of the background random medium. It is essential that the solution depends on these random coefficients in a nonlinear fashion. For instance, the density distribution $\rho(\mathbf{x}) = \exp[-\beta\varphi(\mathbf{x})]$, with $\beta =$ const., is evidently non-Gaussian when the force potential φ is a Gaussian random variable. This distribution cannot be satisfactorily characterized only by the first two statistical moments, the mean value $\langle \rho \rangle$ and the mean square $\langle \rho^2 \rangle$. Moreover, for this distribution the ratios of subsequent moments, $\langle \rho^4 \rangle / \langle \rho^2 \rangle^2$, $\langle \rho^6 \rangle / \langle \rho^2 \rangle^3$, etc., exponentially grow with the moment order. This implies that the spatial distribution is highly non-uniform, the regions of evenly distributed density coexisting with sharp density concentrations.

The question of topological structure of an intermittent distribution is non-trivial because the peaks either form isolated individual "islands" or may be connected into "mountain chains". A standard correlation analysis is ineffective in distinguishing between these two cases. To do this one needs a more delicate method whose possibilities are closer to those of the facility most advanced in pattern recognition, the human ego. The next step in this direction are percolation methods: connection or isolation of given regions that can be determined through investigation of the possibility to travel infinitely along these regions, e.g., with enhanced density.

When the agents that damp higher harmonics are absent (usually when viscosity or molecular diffusivity are neglected), objects with fractal dimension can exist. Variations of a scalar field with time and position and level surface (line) of a scalar field defined in a volume (on a surface) can have different fractal properties (Zeldovich and Sokoloff, 1985).

More versatile features are introduced by vector fields. Curl-free vector fields are essentially gradients of scalar fields. Therefore, more interesting are solenoidal vector fields, velocity field of incompressible fluid for which $\nabla \cdot \mathbf{v} = 0$, and magnetic field among them. In this case, intermittency appears in the form of thin ropes or layers within which the vector field is concentrated. Every rope can be characterized by the flux of the vector across its section. The flux is conserved during motion and stretching of the rope. Toward the end of the rope, where the cross-section varies, even the viscosity does not change the integral flux. A certain fraction of ropes are closed on themselves. The problems of linkages of these ropes and the fraction of the ropes that extend to infinity are also of interest.

Conceptions of intermittency of solutions to linear stochastic differential equations are quite similar to localization as described by solutions of the Schrödinger equation in quantum theory of solids, as introduced and considered by Anderson and Mott (see, e.g., Lifshitz, Gradeskul and Pastur, 1982). This effect was first noticed in the "tight binding model" of Anderson with the lattice Hamiltonian $H = -\Delta + \sigma^2 U(\mathbf{x})$, where σ is the coupling constant and $U(\mathbf{x})$ is the random stationary potential consisting of a sequence of independent identically distributed random quantities. This operator has a point spectrum[c] and corresponding eigenfunctions fall off exponentially (which means localization) in three dimensions only when σ is sufficiently large and for any σ in one and two dimensions. This situation is drastically different from the case of deterministic potentials, typical of ordinary quantum mechanics, where only discrete of continuous spectra are possible. One striking consequence of this theory is a conclusion that, in a one-dimensional medium with even weak disorder, such a transfer process as electric conductivity in weak electric field becomes impossible. These results imply that non-stationary fluctuations in distribution of impurities can play a more important role in conduction mechanisms of real conductors than asserted earlier.

The concept of intermittency was developed independently and even earlier in the theory of hydrodynamic turbulence. Immediately after the formulation of the well-known theory of A.N. Kolmogorov and A.S. Obukhov, it was noted by L.D. Landau that the spectral energy transfer rate ε cannot be constant, and the fluctuations of ε should be important at small scales. Analysis of this possibility has led to a more exact formulation of fundamentals of the theory (see Monin and Yaglom, 1973). It was also found that the distribution of ε at small scales is intermittent (close to the log-normal one); more recent discussion is given by V. Frisch (1985). Moreover, intermittency was also noted in connection with wave propagation in turbulent media (Rytov *et al.*, 1978).

Problems of this type widely use the calculus of path integration.

It is appropriate to compare this approach with classical methods of mathematical physics discussed above. Our predecessors were pleased and reassured by the opportunity of considering a single differential equation instead of an enormous number of irregular trajectories of individual molecules. "One cannot follow all 6×10^{23} molecules in these 20 or more litres of hydrogen." Nowadays we know that in many actual cases, solution of equations for thousands or tens of thousands of "large

[c] We recall that the spectrum is called continuous when every point from some energy interval corresponds to an admissible energy of the system; for discrete spectra there is a finite number of energy levels; for point spectra the number of levels over a finite energy interval is infinite but countable (as, e.g., the set of rational numbers).

particles" becomes accessible with the development of modern computers. In addition, we see again, at a higher level, the relevance of an innumerable number of random paths in grasping such random-related effects as intermittency.

Our discussion of modern problems of the order in random media is based on some classical results of probability theory and statistical physics which are considered in the first chapters. For physicists, randomness is often associated with quantum mechanics, zero vacuum fluctuations or the uncertainly principle. We should stress the differences of our treatment with these cases. The basic equations of quantum mechanics — the Schrödinger, Dirac, and Klein-Gordon equations — are not random equations. Being formulated in appropriate forms, mechanics is quite a deterministic science. Randomness appears only when we endow the wave function with a perspicuous meaning, interpreting it as a probability amplitude for a particle to reside at a given position.

In a complicated picture of interaction between chaos and order, we keep mainly to a new aspect — generation of order by chaos. Another aspect, the onset of chaos from order, has been analyzed for a sufficiently long time and good accounts of it can be found in many books. In particular, the reader is referred to a recent book by Milloni *et al.* (1987).

Thus, we shall spend much time discussing intermittency which is characterized by higher moments of the distribution function. Following the discussion above, intermittency is a son of chance. Nevertheless, in many cases, this offspring has its own importance. Intermittency brings about the problem of geometrical structure of singularities in the random field distribution. Intermittency implies that these regions with exceptionally high values of a vector or scalar make decisive contribution to integral properties even though they occupy a small fraction of a surface or volume.

We clearly realize that our subject is far from being a set of settled problems. Thus, we intentionally make the discussion as qualitative in character as possible in order to make it accessible to a wider audience.

CHAPTER

TWO

THE CHANCE ON STAGE

2.1. CLASSICAL PROBABILITY

All of us are familiar with a common-sense notion of probability. Planning a visit to a physician, a suspicious patient estimates the probability of him having AIDS as, say, 10 percent, which is the average value for the whole population. This can hardly be considered an educated estimate. At best, it estimates the degree of his ignorance or self-delusion. The point is that the 10 percent refers to the average value while every person is interested in the estimate of his personal risk. The latter estimate strongly depends on one's particular way of life, habits, etc. The first scientific concepts of probability were formed with the participation of persons quite reputable, but not very respectable from the modern point of view — gamblers and dicers. Blaise Pascal put an end to the doubts of Chevalier de Meré by proving that two slightly different variants of playing three dice (with the results of eleven or twelve pips) have actually considerably different probabilities. Pascal's idea is utmostly simple in the case of two dice: 11 pips can be realized in two independent cases, $5 + 6$ and $6 + 5$, while 12 pips require the unique combination $6 + 6$. Therefore the probability of having 11 pips is twice that of having 12 pips.

The role of gambling and dicing as a convenient model and basis for the development of the probability concept is indisputable. However, gambling games were not the single source of interest in probability. Trade or insurance can be mentioned, among other fields. This fact is stressed by many historian-scientists, especially by Maistrov (1980). From time immemorial, people pondered about the role of chance in their lives. Following the dissertation of Marx, in this connection philosophers stress the role of Epicurus. It is known that Roman lawyers discussed the average life expectancy of man while Galileo approached the problem of processing experimental data.

Two types of predecessors of the probability theory can be distinguished. The first line of thought discusses general questions, perfecting the conceptual basis before trying to solve any specific problems. It was probably W. von Humboldt (1905) who noted the necessity of this stage of development of science. In this sense, the development of the doctrine of the unity of God the Father, God the Son, and God the Holy Spirit prepared the basis for anticipation of the wave-corpuscular dualism in quantum mechanics. We are more interested in another aspect of the problem: namely, derivation of the first concrete examples. Undoubtedly, de Meré's paradox takes a distinguished place among them. Nowadays gambling is considered an occupation of minor importance, sometimes even with criminal flavour. For the contemporaries of Pascal, it was an important source of initial capital, or the last hope to mend one's finances (cf. Weselovsky, 1909). Genetics, kinetic theory of gases, statistical physics, applied statistics and sociology, and experimental physics became consumers of concrete results of the probability theory much later.

As it often happens, achievements have induced dizziness with success. Along with the healthy development of the probability theory and its applications, many pathological by-products appeared and are still appearing that make popular the joke about three kinds of lie that exist — a common lie, a wicked lie, and statistics. For instance, probability theory was used in an attempt to prove that the Roman Empire never existed (Postnikov and Fomenko, 1980). This was done in the following way. Records of consecutive lengths of reigns of emperors were considered for early and late periods separately and too many coincidences were revealed in these two sequences of numbers. The probability of such coincidence can be shown to be very low. This fact was used to postulate that the historical records were falsified, being a multiple repetition of some shorter records. Combination of a few such excercises opens a possibility to conclude that all Ancient and Medieval history was written in the era of Renaissance. A professional historian, even though he is not a specialist in statistics, notes (Golubtsova and Smirin, 1982) that succession to the throne of the Roman Empire was by no means as regular as, say, the election of new presidents in the United States. Towards the end of the Roman Empire, it was rather a result of complicated interrelations of co-rulers, regional rulers, rebellious colonels, etc. It is not a simple job to determine when one's reign ended and the next began. The imaginary exactness may prove to be an unconscious adjustment to a desired result.

This book is devoted to the role of chance-probability in natural science. Even if we are willing to avoid falling into confusions like what is mentioned above, it is worthwhile recalling the background of the concept of probability.

In gambling games, the probability of winning is understood as a ratio of the number of favourable combinations of cards (or dice positions) to the total number of combinations:

$$P = \frac{n}{N}. \tag{1}$$

This classical concept of probability can be used in other cases when one deals with a finite number of possibilities, all of which are similar in certain sense, i.e., requiring equivalent circumstances which are equiprobable. This concept serves as the basis of the classical probability theory which requires only correct calculation of the numbers n and N, i.e., the ability to count all the possibilities correctly. Remarkably enough, this not very demanding field of thought has obtained numerous non-trivial results.

For example, one's intuition suggests that when two players toss a coin N times, the number of cases when neither player is winning is proportional to N. Actually, it can be shown that the number of draws, i.e., changes of a leader, increases by a random-walk law — it is proportional to $N^{1/2}$ rather than to N (see the textbook on the probability theory by Feller, 1966). After N tosses of a dime, one can either win or lose from 0 to 0.1 $N^{1/2}$ dollars with noticeable probability. Almost always one of the players wins, the draw is only one event from $N^{1/2}$ ones! Therefore, the probability of a draw after N tosses is $\sim N^{-1/2}$. The relative number of the draws, $N^{1/2}/N \sim N^{-1/2}$, declines with N. This can be considered a mathematical proof of the folk-wisdom saying: "Go ahead and play, but do not play to recoup your losses once you fall behind".

Even basic results of the theory are far from widely known. Many "red-and-black" players are still sure that after consequent occurrences of the red field, there is a rather large probability that the next occurrence will be black. This is a delusion: the probability depends only on the mechanism of choosing the winning colour and, after the exclusion of technical imperfections of the Klondike casino described by Jack London, is always equal to 1/2.

We could easily increase the number of examples where the classical notion of probability is applicable, but they turn out to be far from being of interest to modern physics. One of the reasons of this irrelevance is obvious — the finite value of N in classical theory. This is an obstacle, e.g., in the formulation of the theory of experimental data processing where the values errors can acquire belong to a continuous manifold.

Another fundamental restriction is more hidden, behind a modest assumption of equal probability. Which events are actually equally probable? In order to appreciate this problem, we have to consider the probability that all four cards given out from a 52-card pack are aces. The solution is as follows. The total number of equally probable events, i.e., different combinations of four cards, is equal to the number of combinations of 52 elements four at a time, i.e., $C_{52}^4 = 52!/(4!\,48!)$. There is only one favourable outcome — an honest dealer has only four aces in a pack. Thereby, the desired probability is

$$P = \frac{1}{C_{52}^4} \approx 3.5 \times 10^{-6}\,. \tag{2}$$

Now replace the cards by quantum particles that are identical fermions and the combination of four aces is then replaced by the state in which all fermions are in an identical quantum state. Then the Pauli principle dictates that such an event is never possible. Its probability is exactly zero! This simple problem underlines a deep connection between spin and statistics. Probabilities of combinations of boson "cards" must be calculated in their specific way.

Niels Bohr once said that a specialist is a person who is aware of the typical traps and mistakes in his field of knowledge. One of the most widespread mistakes in probability theory is the unjustified assumption of equal probability and independence of events. Another typical mistake is the erroneous application of the probabilistic approach *a posteriori*. For instance, if you are informed that your friend has bought ten lottery tickets and one of them won, please think twice before saying that the probability of payoff in that lottery was about 10%.

Of course, every concrete branch of science that employs the idea of probability can make a discourse upon the notion of probability. This is exemplified by quantum mechanics where the probability amplitude is described by the Schrödinger equation. When this equation is solved, only rudimentary information on probability is needed, and these facts can be easily packed into a few pages of a quantum mechanics textbook. In this very sense, L.D. Landau once said that a physicist does not need special studies of the probability theory. However, the modern science of probability begins where, as a rule, the concept of equal probability cannot be easily applied, and probability distribution cannot be easily obtained by solving certain equations like the Schrödinger equation (for examples see especially Chapter 8).

2.2. PROBABILITY AS A MEASURE

Modern understanding of probability stems from a rather radical idea: consider probability as a measure in the space of elementary events rather than dwell further on the concept of equal probabilities[a] to make it more precise. Even recently this proposition would be understandable only to the high-brow mathematicians. In fact, however, it is a simple and adequate formalization of the concepts routinely used in applications for a long time.

Let us begin with a specific example. Studies of magnetization of rocks have established that the dipole magnetic field of the Earth changes its polarity in random epochs t, t_2, t_3, . . . (Cox, 1968). It seems justified to say that t is a realization of a random variable $t_j(\omega)$, where ω enumerates realizations. From the physical point of view, the reversals are the consequences of some processes that occur within the Earth and which

[a] This idea was proposed by Borel (1909) and realized by Kolmogorov (1936).

are not adequately known (otherwise the values of t_j would be exactly predictable!). Therefore, it is reasonable to consider that ω fills a certain space Ω.

Dividing the time axis into equal intervals of length Δt, we can calculate the probability $P(k)$ that the observed number of reversals k occurs over the period Δt. The number k is also a realization of some random variable which is a function of ω. It is natural to assume that every reversal occurs independently from the previous one and the probability of reversal in period Δt is proportional to Δt. Then

$$P(0) = e^{-\Delta t/\tau}, \quad P(1) = \frac{\Delta t}{\tau} e^{-\Delta t/\tau}, \ldots, \quad P(k) = \frac{(\Delta t/\tau)^k}{k!} e^{-\Delta t/\tau},$$

where τ is the average period of constant polarity. This is the Poisson distribution.

Now we can formulate the relevant concepts in general.

Thus, it is considered that there is a certain space of elementary events Ω whose elements are ω; sets in this space are events. Then the probability of these events is equal to the area or volume occupied by the corresponding set, or, as mathematicians put it, the measure related to the measure of the whole space Ω. As a rule, the unit of area is conveniently chosen such that the measure of the whole Ω is unity.

Even in simplest cases, e.g., in the analysis of experimental errors, we do not know the structure of the space of elementary events, i.e., in fact all the random effects that cause experimental errors. The more so, it is difficult to determine probability over this space.

Nevertheless, it was this concept of probability which turned out to be productive. It proved to be capable of adapting earlier results and of giving new results essential for physics. The reason for this is the fact that over the space of elementary events we can construct certain functions whose properties are practically independent of the structure of this space and the detailed definition of probability. These very properties, with many of which physics is concerned, are the object of the probability theory.

The function $\xi(\omega)$ defined on the probability space is known as a random quantity. Even a function that, say, is equal to seven over the whole probability space is a trivial example of a random quantity. In this sense deterministic quantities are limiting cases of random ones. The value of ξ for fixed ω is called the realization of the random quantity. In every experiment we observe such realizations.

Random quantity is characterized by features that only depend weakly on the unknown properties of the probability space. It is most important that the distribution function $F_\xi(x)$, i.e., the probability that the random quantity ξ takes the values below x:

$$F_\xi(x) = P\{\xi < x\} .$$

If the distribution function is continuous (which is often the case), the variable ξ is called continuous and $F(x) = \int_{-\infty}^{x} p(x)dx$, where $p(x)$ is the probability density. Its physical importance is due to the fact that $p(x)dx$ is the probability of ξ having the value between x and $x + dx$ (the share of the interval dx in the total probability). For non-continuous random variables, the probability density can be considered as a generalized function. We mention that for the Gaussian probability density $F(x)$ is called the error function erf (x) and is tabulated in every reference book on experimental data processing.

Distribution function and probability density are much more tangible than the measure in the probability space. The cost for this tangibility is the incomplete characterization of a random quantity by the probability density. In particular, the same distribution function can characterize entirely different random quantities, e.g., temperature fluctuations and deviations of people's heights from the average value. Such quantities are called identically distributed independent random variables. Formally, independence implies that the probability of ξ_1 having the value between x and $x + dx$ and simultaneously ξ_2 being between y and $y + dy$, which is denoted by $p_{12}(x,y)dxdy$ and called the joint probability density, has the property

$$p_{12}(x,y) = p_1(x)p_2(y) .$$

In other words, probability densities of independent random variables are multiplicative: if you win in tossing a coin with the probability 1/2 and the coin has no memory, the probability of a double payoff is $1/2 \cdot 1/2 = 1/4$.

The probability density is still an excessively detailed characteristic of random quantity. Of course, it can be at least crudely estimated experimentally, but a more realistic goal is to measure the averaged value of random quantity or its moments.

Obviously, $\int_{-\infty}^{\infty} p(x)dx = 1$; that is, any random variable; with unit probability surely acquires some value. The integral can have any value and even diverge. This integral is called the mathematical expectation, or the mean value of the quantity ξ, and is denoted as $M\xi$, or $\langle\xi\rangle$, or $\bar{\xi}$. The physical meaning of the mathematical expectation is described by its jocular description as "an integral of the product of low probabilities by nasty consequences". Returning to thoughts of a physician's patient we note that catarrh is not considered a serious disease although the probability of catching a cold during a year is close to unity; meanwhile, AIDS is considered the "plague of the twentieth century" even though the probability of infection is usually comparatively low — what matters is the value of the "consequences" multiplier.

One can proceed to analyze not only the average but also its powers or other functions. For instance, $\langle f(\xi)\rangle = \int_{-\infty}^{\infty} f(x)p(x)dx$. In particular, $\langle\xi^2\rangle - \langle\xi\rangle^2 \equiv D\xi \equiv \sigma_\xi^2$ is called the variance of the random variable ξ while σ_ξ is known as the standard deviation. Those values of ξ that frequently occur concentrate near the mean value $\bar{\xi}$; the variance, or the standard deviation, characterizes the size of the concentration region. The following useful relation can be easily verified:

$$\langle(\xi - \langle\xi\rangle)^2\rangle = \langle\xi^2\rangle - \langle\xi\rangle^2 = D\xi\,.$$

Considering physical examples, one may encounter random quantities that do not possess mean values because the corresponding integrals diverge. The first example of such a variable was given by Cauchy. Consider a light beam coming from a point at unit distance from a screen. The angle between the beam and the normal to the screen is φ. Consider the situation when φ is randomly distributed over the interval $-\pi/2 \le \varphi \le \pi/2$, i.e., the probability density is given by $p_0(\varphi) = \pi^{-1}$.

The coordinate of the point where the beam reaches the screen, $x = \tan\varphi$, is a random quantity whose probability density decreases so slowly that it has no mean value. Indeed, note that $d\varphi = (1 + x^2)^{-1}dx$ and transform the variable from φ to x in $dP = p_0 d\varphi$. This leads to the following probability density for the coordinate x:

$$p(x) = \frac{1}{\pi}\frac{1}{1 + x^2}; \quad -\infty < x < \infty .$$

We see that $\int_{-\infty}^{\infty} p(x)dx = 1$. However, $\langle x\rangle = \int_{-\infty}^{\infty} xp(x)dx$ diverges (the main value is, of course, zero) and $\langle x^2\rangle = \int_{-\infty}^{\infty} x^2 p(x)dx = \infty$.

However, much more often, one encounters random quantities which have a sufficiently rich set of averaged characteristics, e.g., all power moments $\langle\xi^n\rangle$. The theory of experimental data processing is based on the presumption that experimental errors belong to this class of random quantities.

The mean value of a random quantity usually can be measured experimentally. In order to develop methods of such estimation, the integral $\int_{-\infty}^{\infty} xp(x)dx$ must be considered as an integral in the sense of Lebesgue. In the usual integral calculus, an integral $\int f(x)dx$ is introduced as an area under the curve $f(x)$ (Fig. 2.1). Lebesgue proposed to calculate this area in another way by dividing the ordinate axis rather than the

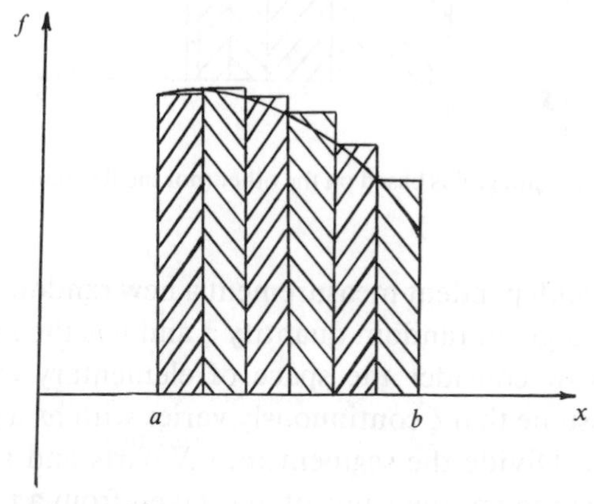

Fig. 2.1. The Riemann integral is understood as the limit of the area shaded in this figure.

abscissa (Fig. 2.2). It would seem that the difference is not critical but now the nature of the variable x becomes unessential, and the only property required is the possibility of introduction of the length along the abscissa — more generally, the measure.

Now, understanding $\int xp(x)dx$, in the sense of Lebesgue, we can say that $\langle\xi\rangle$ is a sum (integral) of the products of ξ by the probability of ξ to have a given value, i.e., $\langle\xi\rangle$ is an integral of ξ over the space of elementary events: $\int \xi d\omega = \langle\xi\rangle$. Here, $d\omega$ is the measure, i.e., the probability in the space of elementary events. From this point of view, the distribution function provides a way to determine a new measure (length) along the axis corresponding to a random quantity. With this new measure, the length of a line segment between the points a and b is given by $F(b) - F(a)$. The average value is now $\langle\xi\rangle = \int x dF_\xi$, where dF_ξ is determined by the measure associated with the distribution function F_ξ.

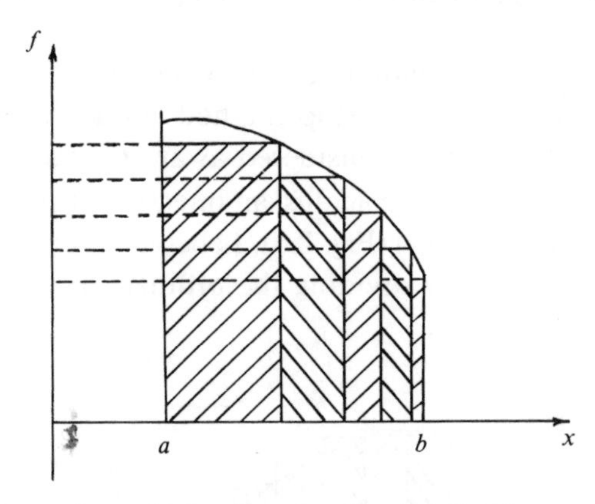

Fig. 2.2. The Lebesgue integral is based on the other (not the Riemann one) construction of the shaded area.

Let us call an independent measurement a new random quantity that is independent of a given random quantity ξ and has the same probability distribution. Now consider the space of elementary events as a unit segment and assume that ξ continuously varies with ω, a point belonging to this segment. Divide the segment into N parts and suppose that for every new measurement the value of ω is taken from a new subsegment.

Then

$$\langle \xi \rangle = \int \xi d\omega \simeq \sum \xi_n(\omega_n) \frac{1}{N} = \frac{1}{N} \sum \xi_n(\omega_n) \,. \tag{3}$$

Here $1/N$ is the measure (probability) of ω belonging to one of the subsegments. Thus,

$$\langle \xi \rangle \simeq \frac{1}{N} \sum_{n=1}^{N} \xi_n \,, \tag{4}$$

where ξ_n are the results of mutually independent measurements. The larger N is, the more accurate is the expression (4). Of course, this derivation is far from rigorous since we presume a specific structure for the space of elementary events, but a more accurate analysis leads to the same result.

Formula (4) is used so frequently that it is often considered a definition of the mathematical expectation. Actually, this understanding is not rigorous and in many branches of physics one encounters intermittent random variables (see Chapter 8), for which the value of $N^{-1}\Sigma\xi_n(\omega_n)$ (often called the empirical average) considerably differs from the true mean value $\langle \xi \rangle$ even for very large N.

Expression (4) can also be understood in another way, invoking the notion of an ensemble of realizations of the random quantity ξ, rather than a sequence of independent measurements. Thus the average value can be understood as the average over the ensemble of realizations.

Expressions (3) and (4) present the essence of the so-called *law of large numbers*. The ideas associated with this law can be developed on the basis of the *central limit theorem* which is a foundation of application of the probability theory in physics.

2.3. THE CENTRAL LIMIT THEOREM

The essence of this theorem is in the statement that, in the limit of large number of summands, not only the equality (4) is true but, in addition, the very sum has the standard (Gaussian) probability distribution which is practically independent of detailed properties of ξ. What matters are the mean value and variance of ξ.

In order to derive the Gaussian distribution, let us find a relation between the probability density of the sum with that of the summands. Let $\zeta = \xi_1 + \xi_2$. For the sum of two independent variables we have

$$F_\zeta(x) = P\{\zeta < x\} = P\{\xi_1 + \xi_2 < x\}$$

$$= \int_{-\infty}^{x} p_\zeta(x')dx' = \iint p(x_1)p(x_2)dx_1\,dx_2\,. \tag{5}$$

The integral is taken over the domain $x_1 + x_2 < x$. Transformation of variables $\tilde{x} = x_1 + x_2$, $\tilde{\tilde{x}} = x_2$ leads to

$$\iint p(x_1)p(x_2)dx_1\,dx_2 = \int_{-\infty}^{x} d\tilde{x} \int_{-\infty}^{\infty} p(\tilde{x} - \tilde{\tilde{x}})d\tilde{\tilde{x}}. \tag{6}$$

Now comparison of (5) and (6) yields

$$p_\zeta(x) = \int_{-\infty}^{\infty} p(y)p(x-y)dy\,.$$

Thus, the probability density of a sum of independent random quantities is given by a convolution of the probability densities of the summands. This is also true for a sum of a large number of random quantities.

A convenient property of convolution is that its Fourier transform is equal to the product of Fourier transforms of individual variables:

$$\tilde{p}_N(k) = \tilde{p}^N(k)\,, \tag{7}$$

where N is the number of variables and the wave marks Fourier transforms.

The Fourier transform of the probability density, i.e., $\int p(x) \exp(ikx)dx$ can be considered the mean value of exp (ikx). This mean value is known as the characteristic function of the random variable ξ. Qualitative properties of the characteristic function of an individual summand, $\tilde{p}(k)$, can be easily understood. Indeed, $\tilde{p}(0)$ is obviously unity. Since the probability density is non-negative, $|\tilde{p}(k)| < 1$ for $k \neq 0$. Let us evaluate the derivatives at $k = 0$:

$$\tilde{p}'(0) = \frac{d}{dk} \int e^{ikx} p(x) dx \bigg|_{k=0} = \int ixe^{ikx} p(x) dx \bigg|_{k=0} = i\langle \xi \rangle ,$$

$$\tilde{p}''(0) = -\langle \xi^2 \rangle, \quad \text{etc} .$$

Thus,

$$\tilde{p}(k) = 1 + ik\langle \xi \rangle - \frac{1}{2}\langle \xi^2 \rangle k^2 + \ldots$$

Now, we are ready to evaluate the characteristic function of the sum. Obviously, $\tilde{p}_N(k) \to 0$ for $N \to \infty$ and $k \neq 0$, since this is the product of N numbers all of which are smaller than unity. Meanwhile, $\tilde{p}_N = 1$ for $k = 0$, i.e., the result is Dirac's delta-function. Its Fourier back-transformation is a uniform distribution. This reflects a widely known fact that any random quantity ξ_n should be normalized by $N^{1/2}$.[b]

For the sake of simplicity put $\langle \xi \rangle = 0$. Non-vanishing average values can be easily accounted for because averages are additive (since the integral is additive).

Dividing every summand by $N^{1/2}$ we see that dispersions of the summands are of order N^{-1}, $\langle \xi^3 \rangle \sim N^{-3/2}$, etc. Therefore, the characteristic function of an individual summand can be represented as

[b] It can be easily verified that normalization by a smaller power of N leads to the same trivial result of the sum of the random quantities increasing with the number of additives. Meanwhile, if it is normalized by a larger power of N, the characteristic function becomes the same unity and the probability distribution is the delta function, $p_N(x) \to \delta(x)$. This also reflects a simple result that a sum of small numbers is small.

$$\tilde{p}(k) = 1 - \frac{\sigma^2 k^2}{2N} + O(N^{-3/2}).$$

Then

$$\tilde{p}_N(k) = \left(1 - \frac{\sigma^2 k^2}{2N} + O(N^{-3/2})\right)^N \rightarrow e^{-\frac{\sigma^2 k^2}{2}},$$

where σ^2 is the dispersion of ξ. We see that the limiting value depends only on the dispersion σ^2. The corresponding probability density is given by

$$p(x) = \frac{1}{\sigma\sqrt{2\pi}} e^{-\frac{x^2}{2\sigma^2}}. \tag{8}$$

This is the well known Gaussian distribution.

We see that a sum of a large number of small independent random quantities has the Gaussian distribution. For the limit $N \rightarrow \infty$, value of the sum of the finite independent random quantities ξ_n can be obtained simply by dividing the above result by $N^{1/2}$:

$$\sum_{n=1}^{N} \xi_n = \sqrt{N}\sigma\zeta, \tag{9}$$

where ζ is the Gaussian random quantity with zero average and unit dispersion. For non-vanishing mean value of ξ_n the right-hand side of (9) is supplemented by an additional term $N\langle\xi\rangle$.

Of course, random quantities are not always independent. Deviation from mutual independence of random quantities ξ_1 and ξ_2 can be characterized by the correlator $\langle\xi_1\xi_2\rangle - \langle\xi_1\rangle\langle\xi_2\rangle$. For independent quantities, this correlator vanishes. There exist dependent random quantities with vanishing correlators of this type, but for Gaussian variables independent and uncorrelated are synonymous.

Inter-correlations of random quantities can strongly affect probabilistic situations. A well known example considers a passenger who goes to an underground station at random by a chosen moment of time. It seems natural that he has equal chances to depart in each direction if he takes the first train that comes to the station. Actually, however, the probability of departure in a certain direction can considerably be larger than 1/2. This is due to the correlation in the schedules of the trains. If, say, the trains in both directions come with the time interval Δt but the train to one direction comes by δt ($\ll \Delta t$) earlier than the other, the probability of taking the train to the latter direction is lower by $\delta t/\Delta t$. Of course, in real life such correlations are not very tight. Nevertheless, the reader can verify this result at the nearest underground station.

Formula (9) expresses the central limit theorem. It is this theorem that has made the probability theory important not only for a *habitué* of the casino but also for a physicist. Experimental data processing heavily relies on the central limit theorem. Of course, every experiment should be treated individually but the majority of methods are based on the central limit theorem.

In a simplest case, the experimenter deals with N independent identically distributed random quantities. Their realizations are the results of measurements. These random quantities have the same mean value $\langle \xi \rangle$ whose determination is the goal of an experiment. Root-mean-square deviation σ of these quantities is usually known and it determines the accuracy of the measurements. Arithmetic mean of the measured values is a random variable with the same mean value $\langle \xi \rangle$ and the root-mean-square deviation $\sigma N^{-1/2}$. According to the central limit theorem, for large number of measurements N, the arithmetic mean is approximately a Gaussian distribution whose dispersion decreases with N. Estimation of the difference between the arithmetic mean of measured values and the true mean value $\langle \xi \rangle$ requires only a standard procedure available in any modern computer as the least squares algorithm. The problem is to choose a relevant procedure. In order to develop this knowledge, consider a specific example of data processing.

2.4. THE LEAST SQUARES METHOD

Our example considers the problem of determination of the age of rocks with the use of physical methods. Practically every rock contains, in microscopic quantities, radioactive elements. Relations among abundances of a given radioactive element and products of its decay allows to calculate the age of the rock. One concrete method is based on the analysis of abundances of the radioactive isotope of strontium ^{87}Sr and a product of its decay ^{87}Rb. The laws of radioactive decay imply that, provided the sample under investigation has been an isolated system whose parts have not exchanged its matter during its lifetime, the measured points lie on a straight line on the plane $x = {}^{87}\mathrm{Rb}/{}^{86}\mathrm{Sr}$ and $y = {}^{87}\mathrm{Sr}/{}^{86}\mathrm{Sr}$. Slope of this line is essentially the sample age, therefore the line is called an isochrone, and the abovementioned plane, the isochrone plane. Thus, determination of the sample age reduces to the problem of drawing a straight line, the isochrone, based on data on abundances of the isotopes of strontium and rubidium, that are, of course, contaminated by errors.

The very errors of measurements are also determined experimentally and their values are also some random variables, more or less deviated from the true values. In a common situation, measured abundances and their ratios have small errors, but the values of the errors can be estimated only crudely. For instance, the strontium-rubidium abundance ratio can be measured with accuracy to the order of 2%, but it cannot be said whether it is 1.5%, 1.7% or even 2.3%. In this case it is prudent to somehow overestimate the error, putting it equal not to 2%, but, say, to 3%, as a guaranteed upper limit.

This presumption allows to obtain a simple measure of the scatter of measured points around a straight line. Let the slope a and some initial value of the abundance ratio b be given. It is natural to measure x and y in units of σ_x and σ_y, respectively. For given standard errors σ_x and σ_y, a point can wander from the line to the largest possible distance if it deviates along the direction orthogonal to the line. This distance can be easily calculated also in the original coordinates. The sum of these squared distances is that very measure of the scatter of the point about the line. Of course, it depends on the points set and on a and b. The method of the least squares consists of the minimization of the normalized sum of squared deviations, the so-called residual. In the considered example it is given by (Bruks *et al.*, 1972)

$$S(a, b) = \sum_{i=1}^{n} \frac{(y_i - ax_i - b)^2}{\sigma_{x_i}^2 a^2 + \sigma_{y_i}^2}\ .$$

On the other hand, we can estimate the residual from the point of view of the probability theory, considering it as a sum of n squared Gaussian quantities. The quantity, which serves as an upper admissible value of the residual $S(a, b)$ and corresponds to taking into account only those events whose probability is at least $1 - P$ (say, $P = 0.95$, or 95%), is denoted by $\chi^2_{n-2}(P)$. Its value can be found in tables of mathematical statistics. For $n > 30$, the following asymptotic expressions can be used:

$$\chi^2_n(P) = \frac{1}{2}\left(\sqrt{2n-1} + u_p\right)^2,$$

or

$$\chi^2_n(P) = n\left(1 - \frac{2}{9n} + u_p\sqrt{\frac{2}{9n}}\right)^2, \tag{10}$$

when P is close to 0 or 1, respectively. The quantity u_p (inverse error function) indicates how many standard deviations corresponds to the adopted reliability level for the normally distributed random variable. For instance, $u_p = 1.9600$ for $P = 0.95$ while $u_P = 3.2905$ for $P = 0.999$. Asymptotic expressions (10) imply that, according to the central limit theorem, for large n the sum of squared deviations has a normal distribution.

For an adopted value of P only those values of a and b are reliable, for which

$$S(a, b) \leqslant \chi^2_{n-2}(P). \tag{11}$$

Before proceeding to the solution of inequality (11), which is the central point of the method under discussion, we should make a few notes. Inequality (11) is justified based on a number of presumptions. First, errors are considered independent along the x- and y-axis. This presumption can be verified by measurements of abundances and isotope ratios in independent experiments. Furthermore, in the calculation of $\chi^2_n(P)$, the errors are assumed to be Gaussian. In other words, it is believed that experimental errors are formed under the influence of many factors of nearly equal importance, rather than by some single factor. In principle, this can also be verified but, when there are many measured points, deviations from Gaussian statistics are negligible, again due to the central limit theorem. Dealing with small number of measurements is also dangerous because, strictly speaking, expression (11) is applicable only when the function $S(a, b)$ is either quadratic in a and b (it can be verified that this is true only when the errors along one of the axes are essential, i.e., either $\sigma_x a \gg \sigma_y$ or $\sigma_y \gg \sigma_x a$), or for large number n of measured points. Experience in dealing with statistical measurements indicates that n greater than 30 is very good (which is also implied by a good applicability of the asymptotic expressions for $\chi_n^2(P)$ for these values of n). An admissible value is $n \simeq 10$. In practice, one often meets even smaller values, $n \simeq 5$–7; in these cases statistical results should be treated with caution. And when $n = 3$–4, it is wiser to draw a straight line with a ruler and estimate the errors simply by inspecting the plot.

Inequality (11) is rather complicated and its straightforward solution requires serious computations. Very frequently, experimental errors are not very large, and parameters a and b can be successively found with a ruler. This simple procedure can often yield the values a^* and b^* corresponding to the minimum of the residual with an accuracy of 1 percent or even better. It is expected that true values of a and b deviate from a^* and b^* not very strongly, by a few tenths of a percent. Formally, in inequality (11), the unknowns are a and b, but in the denominator of the residual a is multiplied by σ_x, the accuracy of the latter quantity being much inferior to that of a. In this case it is natural to replace (11) by the approximate inequality

$$\tilde{S}(a, b) = \sum_{i=1}^{n} \frac{(y_i - ax_i - b)^2}{\sigma_{x_i}^2 A^2 + \sigma_{y_i}^2} \leqslant \chi_{n-2}^2(P)\,, \tag{12}$$

where A is the value of a estimated by the eye that corresponds to the minimum of $\tilde{S}(a, b)$. Solving inequality (12) instead of (11) we introduce an additional error which is negligible in the considered case.

Inequality (12) can be solved quite easily. Level lines of the function $\tilde{S}(a, b)$, i.e., the lines $\tilde{S}(a, b) = \text{const.}$, on the (a, b)-plane (not on the isochrone plane) are concentric ellipses with centers at the point $\bar{a}$, $\bar{b}$ where $\tilde{S}(a, b)$ reaches a minimum. The values of $\bar{a}$ and $\bar{b}$ can be found from

$$\frac{\partial \tilde{S}}{\partial a} = \frac{\partial \tilde{S}}{\partial b} = 0$$

to be

$$\bar{a} = \frac{\alpha_{11}\alpha_{23} - \alpha_{13}\alpha_{22}}{\alpha_{11}\alpha_{22} - (\alpha_{12})^2}; \quad \bar{b} = \frac{\alpha_{12}\alpha_{13} - \alpha_{11}\alpha_{23}}{\alpha_{11}\alpha_{22} - (\alpha_{12})^2}$$

where

$$\alpha_{11} = \sum_{i=1}^{n} \frac{x_i^2}{A^2\sigma_{x_i}^2 + \sigma_{y_i}^2}; \quad \alpha_{22} = \sum_{i=1}^{n} \frac{1}{A^2\sigma_{x_i}^2 + \sigma_{y_i}^2};$$

$$\alpha_{12} = \sum_{i=1}^{n} \frac{x_i}{A^2\sigma_{x_i}^2 + \sigma_{y_i}^2}; \quad \alpha_{13} = -\sum_{i=1}^{n} \frac{x_i y_i}{A^2\sigma_{x_i}^2 + \sigma_{y_i}^2};$$

$$\alpha_{23} = -\sum_{i=1}^{n} \frac{y_i}{A^2\sigma_{x_i}^2 + \sigma_{y_i}^2}.$$

If $S_{\min} = \tilde{S}(a^*; b^*)$ is greater than $\chi^2_{n-2}(P)$, then there are no values of a and b that satisfy (12) and the very isochrone model must be rejected. If, on the contrary, $\tilde{S}(\bar{a}, \bar{b}) \leqslant \chi^2_{n-2}(P)$, then we are able to draw an ellipse which envelopes the confidence region, i.e., the set of solutions to inequality (12) that represent admissible values of the age and initial isotope ratio.

One can easily write out the parameters of this confidence ellipse, e.g., lengths of its semi-axes. But they are not easy to interpret and do not give directly estimates like $\bar{a} \pm \delta a$ and $\bar{b} \pm \delta b$ which are usually published.[c]

[c] Note that the situation is even worse for the more rigorous inequality (11); in this case, the confidence region is not an ellipse but a more complicated figure.

This difficulty is resolved in the following way. Instead of indicating the boundaries of the confidence region, the values of errors for a and b are published which only characterize crudely the size of the confidence region. Fig. 2.3 illustrates this situation. The values of δa and δb can be estimated from the simple expressions

$$\delta a = \sqrt{\frac{\chi^2_{n-2}(P) - S_{\min}}{\alpha_{11}}}; \quad \delta b = \sqrt{\frac{\chi^2_{n-2}(P) - S_{\min}}{\alpha_{22}}}.$$

The values $\overline{\delta a}$ and $\overline{\delta b}$ (see Fig. 2.3) are somewhat large. It can be easily seen that

$$\frac{\overline{\delta a}}{\delta a} = \frac{\overline{\delta b}}{\delta b} = \left(\frac{\gamma^2}{1-\gamma^2} + \sqrt{1-\gamma^2}\right),$$

where

$$\gamma = \frac{\alpha_{12}}{\sqrt{\alpha_{11}\alpha_{22}}}.$$

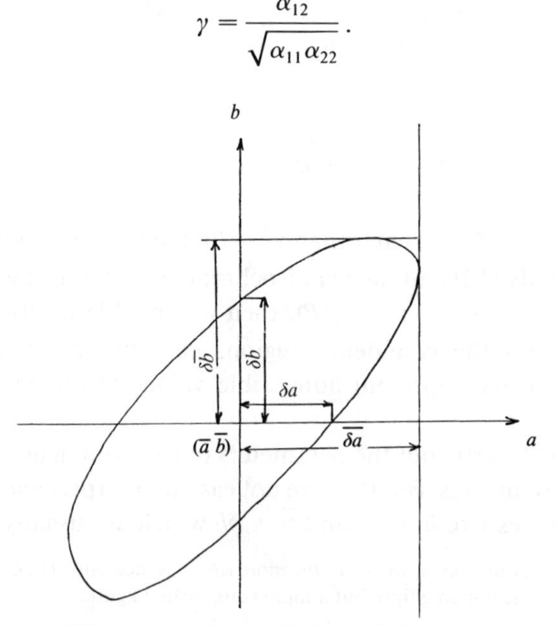

Fig. 2.3. Different types of uncertainties in the least squares method.

It is clear that the estimates of the errors cited in the literature are necessarily very crude and it is hardly reasonable to try to estimate them with accuracy of percentage.

We should also note that the confidence set and, consequently, the errors δa and δb are random values themselves. This means, say, that if we have determined the errors with reliability $P =$ 95% and 20 isochrone lines should be plotted, one of them can turn out to be erroneous. Of course, it can even be the first plot: errors are always unexpected.

Let us illustrate this method using the studies of the Kalar range as an example (Vinogradov *et al.*, 1983; Sokoloff and Bujakaite, 1984). Experimental data are given in Fig. 2.4. The bars correspond to one standard error along x and y axes. Undoubtedly, the points crowd around the straight line. However, the minimal possible residual is $S_{\min} = 50.3$, which considerably exceeds χ_9^2 (0.95) = 16.9. This means that a single straight line cannot be drawn through all the points. Notice now that experimental points cluster in two distinct groups. For each group a straight line can now be drawn confidently. For the first group, $a = 0.04 \pm 0.001$, $b = 0.071 \pm 0.0003$, $S_{\min} = 5.5 < \chi_4^2(0.95) = 9.5$. For another group $a = 0.037 \pm 0.001$, $b = 0.7008 \pm 0.0004$, $S_{\min} = 0.94 < \chi_3^2(0.95) = 7.8$. In both cases $\gamma \approx 0.6$ and, therefore, $\overline{\delta a} = 1.4\ \delta a$ and $\overline{\delta b} = 1.4\ \delta b$.

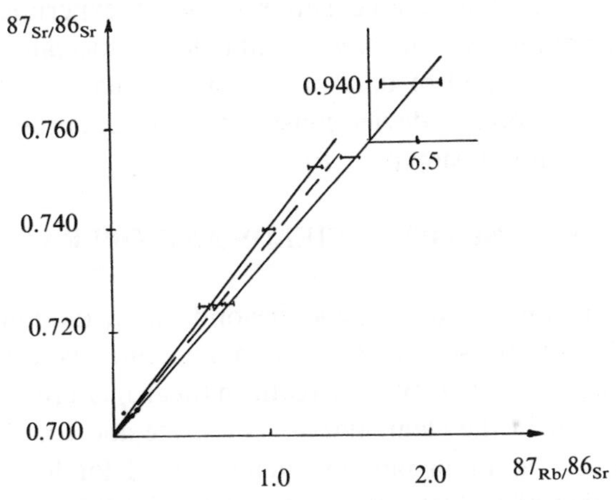

Fig. 2.4. Experimental points are poorly approximated by a single line (dashed) but a pair of straight lines (solid) provides a good fit.

Thus, the considered data indicate that there occured two events in the history of the Kalary range, with ages 2.80 and 2.55 billion years.

We considered only one variant of experimental data processing with least-squares method. The characteristic feature of this method is that the resulting estimates are unbiased. In other words, when this estimate is considered as a random variable, its mean value is exactly equal to the true value. Meanwhile, the errors of estimated parameters decrease very slowly with the growth of the body of experimental data. In expressions for the errors (13), denominators α_{11} and α_{22} are of the order of the number of measured points n; the quantities χ^2 and $S_{\min}$ in the nominator are also of the order n. However, for a correct estimation of accuracy of an individual measurement, it can be deduced from (10) that the difference $\chi^2 - S_{\min}$ is of the order of $\sqrt{n}$, i.e., $\delta a \propto n^{-1/4}$ and $\delta b \propto n^{-1/4}$. When the error of an individual measurement, σ, is significantly overestimated, the values of δa and δb do not decrease with the growth of n.

In the examples discussed above, we were concerned mainly with the verification of the applicability of the model rather than the most accurate estimation of the model parameters a and b. If one has a large number of measured points (n is very large) and is confident in the applicability of the model, then accuracy of estimation of the model parameters can be considerably improved at the price of abandoning the unbiased nature of the estimate. Small bias in the estimate sometimes allows to reach a considerable suppression of experimental errors. Moreover, in certain cases, an experimental device specially coordinated with the applied method of data processing can have a better performance than a more precise device constructed without proper specialization (see Pytyev and Shishmarev, 1983).

2.5. RANDOM FUNCTIONS, FIELDS AND OPERATORS

The description of nature usually goes beyond simple random quantities. The physical characteristics of objects vary in time as, e.g., an output noise of an audio amplifier. What is realized (heard) is a function $\xi(t, \omega)$. Here ω is a point in the elementary-event space and t is the time. For fixed t, $\xi(\omega)$ is a common random quantity, and for fixed ω, it is an ordinary function of t which called a realization of the random function $\xi(t,\omega)$. Random functions of time are usually called *random processes*.

As we discussed in Section 2.2, probability can be understood as a measure in the space of elementary events. Random quantity determines a probabilistic measure on a line,

$$d\mu = p(x)dx = dF(x)$$

(this measure is known as the Stiltjes measure), where x is the value of the random variable, $p(x)$ is the associated probability density. A random process determines the measure in the functional space. This approach will be discussed in the next chapter on the example of the Brownian motion.

It is natural that for two very different moments of time t_1 and t_2, random quantities $\xi(t_1)$ and $\xi(t_2)$ can be considered independent with great accuracy. In other words, random processes do not remember their past. Of course, formally one can envisage a random process with a good memory, but usually such cases are rather rare. For close moments, the quantities $\xi(t_1)$ and $\xi(t_2)$ are not independent: any physical system has at least a small inertia and cannot forget its past immediately. Time is a dimensional quantity and the difference $|t_1 - t_2|$ can be large or small only in comparison with a certain characteristic time τ. The characteristic time is individual for every random function and is known as the correlation time. Just like any other characteristic parameter of a physical system, the notion of characteristic time is somewhat ambiguous and its definition should be stated in every single case. Different definitions can lead to the values that differ by a multiplier of the order of unity.

The degree of mutual dependence of the values of a random function is characterized by the (auto) correlation function

$$B(t_1, t_2) = \langle \xi(t_1)\xi(t_2) \rangle - \langle \xi(t_1) \rangle \langle \xi(t_2) \rangle ,$$

that vanishes for independent functions

$$\langle \xi(t_1)\xi(t_2) \rangle = \langle \xi(t_1) \rangle \langle \xi(t_2) \rangle .$$

When a random function $\xi(t)$ is statistically homogeneous in time (stationary), the correlation function depends only on the time difference $B = B(|t_1 - t_2|)$. Note that individual realizations $\xi(t)$ are, of course, non-stationary.

It is commonly presumed that correlations decay exponentially, as $\exp(-|t_1 - t_2|/\tau)$ or even as $\exp[-(t_1 - t_2)^2/\tau^2]$, for $|t_1 - t_2| \to \infty$. However, cases of slower, power-law decay can also be met (Bunimovich and Sinai, 1981).

When a random function is a sum of many independent random contributions, the central limit theorem states that at any moment it has the Gaussian distribution. The Gaussian random function can be completely described by its mean value and its correlation function. In particular, its dispersion is equal to $B(t, t)$. The non-Gaussian random processes can, in principle, be described by a set of more complicated correlations:

$$\langle \xi(t_1)\xi(t_2)\xi(t_3)\rangle, \quad \langle \xi(t_1)\xi(t_2)\xi(t_3)\xi(t_4)\rangle,$$

and others, that now cannot be expressed through the mean value and the correlation function.

Random fields that vary in space, $\xi(\mathbf{x}, \omega)$, e.g., the random temperature field, can be introduced in a similar way. Such fields are characterized by a spatial correlator

$$B(x, y) = \langle \xi(\mathbf{x})\xi(\mathbf{y})\rangle - \langle \xi(\mathbf{x})\rangle \langle \xi(\mathbf{y})\rangle$$

which depends only on $r = |\mathbf{x} - \mathbf{y}|$ in the case of statistical spatial homogeneity and isotopy. For small r, the correlation function is large; for large ones, it is small. The characteristic distance r_0, which determines the smallness of r, is called the correlation radius. A frequently used approximation is $B(r) \propto \exp(-r/r_0)$ or $\exp(-r^2/r_0^2)$. It is convenient to deal with dimensionless correlation functions for which $B(0) = 1$.

Random fields can be considered not only in a physical coordinate space. In many circumstances, the Fourier representation proves to be useful. In this case, the spatial correlator transforms to

$$B(\mathbf{k}_1, \mathbf{k}_2) = \langle \xi(\mathbf{k}_1, \omega)\xi(\mathbf{k}_2, \omega)\rangle - \langle \xi(\mathbf{k}_1, \omega)\rangle \langle \xi(\mathbf{k}_2, \omega)\rangle ,$$

where $\xi(\mathbf{k}, \omega)$ is the Fourier transform of $\xi(\mathbf{x}, \omega)$.

Spatial homogeneity implies that

$$B(\mathbf{k}_1, \mathbf{k}_2) = B(\mathbf{k}_1)\,\delta(\mathbf{k}_1 + \mathbf{k}_2) ,$$

where the function $B(\mathbf{k})$, called the power spectrum of the random field, is the Fourier transform of $B(r)$.

If a random field depends both on space and time, it has both correlation time and correlation radius; its correlation function depends both on position and time. Vectorial and tensorial random fields are described by correlation tensors.

A field of an isotropic and homogeneous scalar random quantity (say, temperature) can easily be described, e.g., by its Fourier-transform. This is accomplished by the prescription of the power spectrum of the field as a squared function of the field depending on the frequency and modulus of the wave vector. Meanwhile, the representation of a random vector essentially depends on its nature. A random curl-free field can be represented as a gradient of a random scalar function, and then the problem of description reduces to that of a scalar field. For the solenoidal random field (random magnetic field or turbulent velocity of incompressible fluid), its three components cannot be considered as independent random scalar fields, because the divergence-free condition must be satisfied. This condition can be expressed through the field correlators (Batchelor, 1953; Monin and Yaglom, 1973). However, the solenoidality can be taken into account through the structure of the field itself. This requires introduction of the special quantities designed for the description of a random solenoidal field. To be concrete, we shall refer to the magnetic field.

Let us introduce a set of surfaces $\eta(r) =$ const., with the field of normal vectors $\mathbf{n} = \nabla\eta$. Then magnetic field can be introduced with the help of two scalar potentials Φ and Ψ:

$$\mathbf{H} = \nabla \times (\Phi\mathbf{n} + \Psi\nabla \times \mathbf{n}).$$

Thus, we see that the random magnetic field $\mathbf{H}(\mathbf{x}, \omega)$ is described by the three random scalar fields $\eta(\mathbf{x}, \omega)$, $\Phi(\mathbf{x}, \omega)$ and $\Psi(\mathbf{x}, \omega)$ (see Zeldovich et al., 1985). This description does not impose any symmetries and ensures statistical homogeneity and isotropy of the field $\mathbf{H}$, provided the fields η, Φ, and Ψ are statistically homogeneous. The solenoidality is ensured automatically.

The theory also considers random operators, e.g., the heat transfer operator $(\mathbf{v}\cdot\nabla) + \kappa\Delta$, where the velocity field $\mathbf{v}(t, \mathbf{x}, \omega)$ is random. The diffusivity κ can also be a random quantity. Random operators can be introduced in a more general way, but here we consider only the so-called operators with random coefficients.

Fortunately, the introduction of these simplest operators does not require the development of new approaches to the solution of associated differential equations to prove the existence and uniqueness of theorems, etc. It is sufficient to recall that, for a fixed random parameter ω, an operator with random coefficients is a common deterministic operator. Randomness only requires the introduction of the probabilistic measure in the operator space or the space of solutions of associated differential equations.

2.6. ERGODIC PROPERTIES, CHAOS, INTERMITTENCY

The central limit theorem has made a great impact on physicists and mathematicians, which is indicated even by its name. It has enabled mathematics to be applied to the studies of chaos in nature. The physical concept has formulated chaos as a state that arises from the initial order due to many mutually independent influences. According to the central limit theorem, this state must be described, from probabilistic viewpoint, by the Gaussian distribution. Depending on the tensorial properties of the quantity under investigation, one would consider the Gaussian random quantity, the random scalar and vector fields, etc.

The concept of chaos is closely linked with another concept of physical statistics, i.e., ergodicity. Averaged values can be introduced not only as integrals over the space of elementary events but also as averages over ensemble, i.e., over many realizations of a random quantity. When one considers a random field where correlations decay in space and time, i.e., are of a statistical ensemble, it can be delegated to many considered space-time points. Indeed, values of a random field and distant points are practically independent and reproduce the result of independent realizations. Thus, the ergodic hypothesis asserts that

$$\langle \xi(t, \mathbf{x}) \rangle = \frac{1}{V} \int \xi(t, \mathbf{x}) d^3\mathbf{x} \,,$$

or

$$\langle \xi(t, \mathbf{x}) \rangle = \frac{1}{T} \int \xi(t, \mathbf{x}) dt \,,$$

the ensemble average is equal to both the spatial average and time-average. Formally, this assertion implies averaging over infinite space-time but, physically, one always takes the averages over the physically large space and time domains whose sizes far exceed the characteristic scales of the problem at hand. Therefore, for a physicist ergodicity means absence of large-scale structures which can strongly affect the spatial or time average even when they are widely spaced.

Classical thermodynamics considers the strange and inexplicable spontaneous formation of structures in nature, for example, the patterns of frost on window panes during the cold winter, the regular arrays of convective cells in liquid, or the coloured layers in the gems jasper and agate. Any sort of lining matter seems to violate thermodynamics. It can be easily understood how the entropy can decrease in its self-organization processes. Increase in the entropy of the system as a whole, e.g., with regard to the heat transfer during ice crystallization or, during the food combustion in a living organism, is much greater than the local decrease of entropy associated with the formation of the structures.

A principally different approach invokes the chance to explain the spontaneous order in many particle systems. The approach can be compared with synergetics which employ somewhat similar ideas. However, the cornerstone of synergetics is of nonlinear processes and chance is only invoked to produce a small seed for nonlinear instabilities.

Meanwhile, structures are typical of random media in transfer phenomena that are described, at a certain stage, by linear differential equations. In these cases, randomness is a basic mechanism of creation of a structure, while nonlinearity, which interferes at the later stages, prevents its infinite growth. The structures that thereby arise in random media have a very special character: they are essentially strong enhancements, or peaks randomly scattered in space and time. The gaps between the peaks of the random quantity has low values; the gaps themselves are widely spaced and rare.

A general term that describes such a picture is "intermittency". This concept has been introduced by Batchelor and Townsend (1959) for the patched temperature distribution in turbulence. Intermittency was also studied in hydrodynamic turbulence in connection with Landau's refinement to the hypothesis of Kolmogorov and Obukhov (see, e.g., Monin and Yaglom, 1973), and in the theory of wave propagation in random media (Rytov *et al.*, 1978). One more example is the localization effect the in the quantum theory of disordered media thoroughly studied by I.M. Lifshitz and his co-workers (1982).

Intermittency has been revealed by numerical experiments in magnetohydrodynamics (Meneguzi *et al.*, 1981) as well as in the galaxy formation theory that has been supported by astronomical observations of the structure of the universe (Shandarin and Zeldovich, 1986).

From a physical point of view, intermittency arises due to the coherence introduced by the random medium into the process of transfer of the considered physical quantity. For example, in a random flow of conducting fluid with an embedded magnetic field, there exist places where the flow efficiently amplifies the magnetic field by stretching the field lines (Molchanov *et al.*, 1985). Of course, such places are rare, having a small chance to happen. However, these maxima make a dominating contribution to the magnetic energy. They determine the average field and its square and cannot be ignored. The two first moments, however, are not enough to describe the magnetic field distribution completely. The principal and characteristic property of intermittency is the abnormal relation (when compared with the Gaussian one) between the consequent statistical moments. In terms of the Fourier analysis, intermittency is characterized not only by the slow decrease of amplitudes of the Fourier modes with wave number, but also by certain phase relations among them. If instead the high modes had random phases, their combination would be close to the Weierstrass function, i.e., a fractal distribution rather than isolated peaks. On the contrary, a sum of specially phased plane waves are combined into a delta-function or a set of delta-functions.

In a statistically homogeneous medium, intermittency can be a pronounced phenomenon: in the presence of instability, e.g., when the magnetic field is self-excited, the ratio of the mean magnetic energies in the regions occupied by the peaks of the magnetic field and the background regions grows exponentially in time. On the other hand, in a spatially bounded domain, the characteristic distance between the high peaks grows, and at a certain moment of time, it becomes greater than the size of the domain. Afterwards, the spatial and ensemble averages cease to be identical, with the spatial averages growing slower than the ensemble ones. This intermittency in a bounded problem can be less sharply expressed.

The spatial structure of an intermittent distribution is a very important property. Temporal evolution asymptotically brings to high isolated peaks; in space, these peaks correspond to highly magnetized spots divided by vast regions of weaker intensity. However, an intermediate asymptotic behaviour may be, or even typically is, the formation of cellular or network structure: thin channels of high intensity ("enriched phase") isolate, one from another, islands of "poor phase". An example of such intermediate-stage intermittency is provided by the structure of the universe (Shandarin and Zeldovich, 1984). In optics, the problem of structures arises when one considers the passage of light through a plate with a random profile. This situation is described formally as an evolution of a random initial distribution in a deterministic medium. At a short distance from the random surface, the structure determined by caustics reappears. It is washed out at greater distances from the plate; and the influence of initial random conditions diminishes.

These intermittent phenomena can be directly observed (Fig. 2.5) or revealed numerically (Fig. 2.6).

Note that the problem of the onset of intermittency is, in a certain sense, inverse with respect to the widely discussed problem of the onset of chaos from regular motion in a dynamical system governed by a small number of ordinary differential equations.

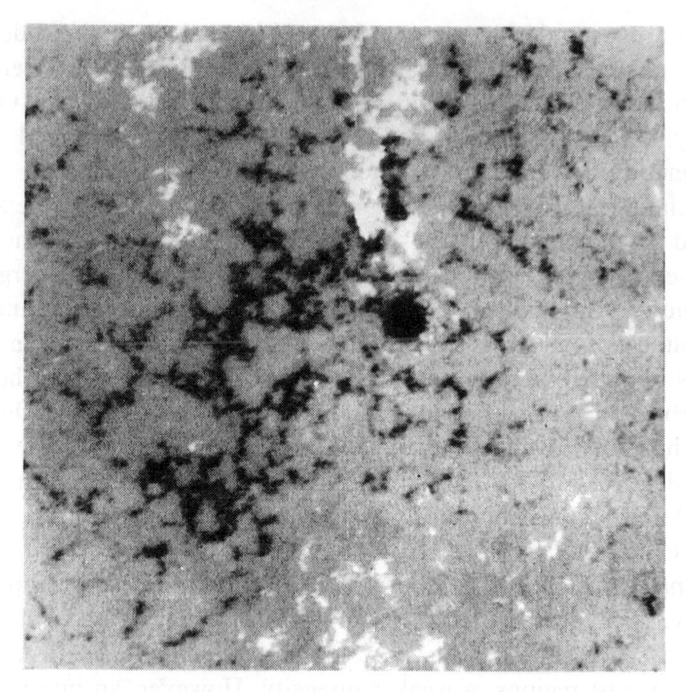

Fig. 2.5. A magnetogram obtained at the observatory Kitt Peak in 1982 (Giovanelli, 1984).

Even a Gaussian random field is not free from some elements of order. The general background of a Gaussian field is produced by a random, unordered variation of the quantity φ whose amplitude is of the order of $b^{1/2}(0)$. The rare high peaks can be detected against this background. In general, the level surfaces near a peak are close to triaxial ellipsoids. However, the higher a peak is, i.e., the larger the ratio of its height to the r.m.s. deviation, the more probable is the mutual closeness of all these axes, the closer are the level surfaces to sphere. Therefore, near the top, a peak is not sharp but has the regular shape of a rotationally-symmetric surface whose meridian in an isotropic case, which is similar to the plot of the function $b(r)$, $r = |\mathbf{x} - \mathbf{y}|$. One can easily estimate the maximum peak height in a given sphere and the distance between the peaks.

Of course, these peaks can also be detected when one observes many realizations of a Gaussian quantity. The existence of the peaks is widely known but they play a minor role in the usual operations with the

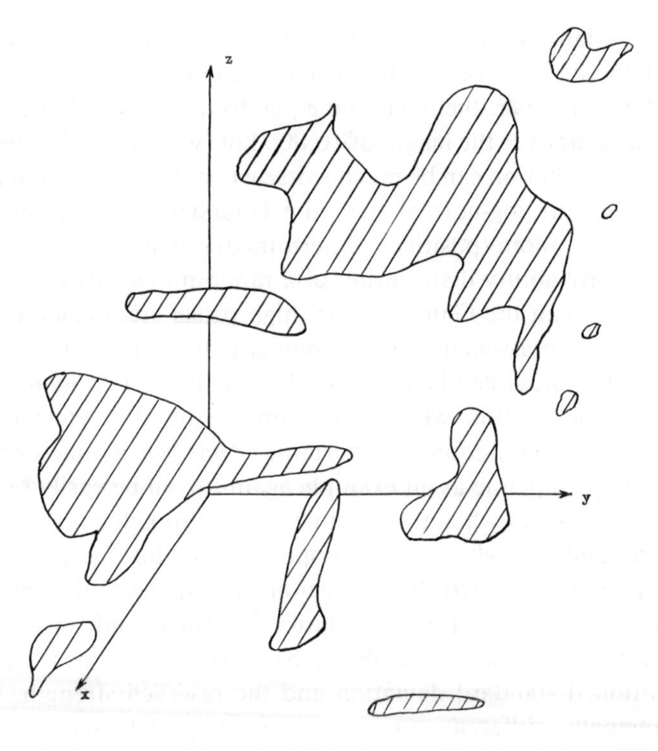

Fig. 2.6. Magnetic concentrations (shaded) in the numerical simulations of Meneguzzi *et al.* (1981).

Gaussian random quantities, e.g., in processing of experimental data with the least square method. Roughly speaking, one can ignore the peaks of arbitrary heights by the consideration that the Gaussian quantity has a maximum deviation equal to three standard deviations (the Gauss rule). However, in many cases, such an approach can lead to dramatic errors. Consider for example, a very long pipeline whose breaking strength is uniform with high accuracy, but which is afflicted with a small random Gaussian error. When the standard deviation of the strength is much smaller than the reserved strength of the pipeline, one is tempted to believe that the pipeline would be very reliable, i.e., it has a low probability to rupture. This logic is typical of the existent pipeline designs. However, this assertion is true only for short pipelines. The breakdown of a very long pipeline can be caused by a single, improbable damage. This damage can be caused by a very rare deviation that is much larger than the standard one. If the correlation length along the pipeline is in the order of a meter, the deviation equal to ten standard ones can be expected to occur over the length of 10,000 km (see similar discussion in Sect. 7.5). The situation can be much worse when the distribution is non-Gaussian, e.g., the log-normal with the Gaussian distribution of the logarithm of a random (positively determined) quantity.

When the probability distribution of a random quantity decreases at infinity slower than the Gaussian one, high peaks are, evidently, more abundant, i.e., the associated structureness is more pronounced. Such enhancement of order can be intrinsically associated with many effects; the clearest is the situation when the random deviation is determined not by many independent, equally important, effects but by one dominant factor. Taking a pipeline as an example again, it can be easily seen that such a situation arises when the pipeline passes through a region of land with rapidly and strongly variable properties. For instance, it can be a region of permafrost where frozen regions are mixed with thawed-out ones. The exploitation of pipelines shows that such conditions are very unfavourable for estimating reliability based on the comparison of the abovementioned standard deviation and the reserved strength. In this case, principally different arguments are required, such as those employed in the studies of the conductivity of disordered metallic media.

When a random error that afflicts the measurements has a non-Gaussian tail of probability distribution, the application of the least square method can lead to serious mistakes. It follows, in particular, that one should single out the results spoiled by the wild errors in measurements. Such wild errors have nothing in common with the Gaussian random errors that occur regularly and the wild error, of course, cannot be averaged together with the normal ones. The selection of faulty results does not represent a problem when the data set is not very large and an experimenter can control the results. However, the situation is more complicated when the data processing must be completely computerized. This requires the development of special methods insensible to faults in measurement. Such methods are known as *robust* (see, e.g., Dawldy and Eckman, 1984).

It is not difficult to give an example of probability distribution that gives less pronounced peaks than the Gaussian distribution. Such distribution has the quantity $\zeta = \sigma\varphi^2/(\sigma^2 + \varphi^2)$, where φ has a Gaussian distribution. It is evident that ζ simply cannot have any peak whose height considerably exceeds the standard deviation.

2.7. EQUILIBRIUM AND TRANSITION PROBABILITY

The concluding two sections of this chapter are devoted to the concept of probability, considering a typical problem of establishment of a stable equilibrium.

This can be a chemical equilibrium: the considered substance A is formed from another substance B and can, in turn, be decomposed producing B again. The equation that governs the number of molecules of the substance in a vessel A, denoted by n, is given by

$$\frac{dn}{dt} = k_1 n_B - kn\,.$$

The coefficients k_1 and k are given constants. The concentration of the substance B, n_B, is considered large and practically constant during the reaction, so that the product $k_1 B$ can be considered constant and denoted as b. Thus

$$\frac{dn}{dt} = b - kn\,. \tag{13}$$

A similar equation also describes the following situation. Imagine a vessel with a small hole that is connected with a large vessel (the atmosphere). In the large vessel the gas concentration n_0 is constant. The problem is to calculate the number of the density of molecules in a small vessel. The smallness of the hole, as compared with the free path of the molecules, is essential for the flow of the gas to have the character of random passage of molecules through the hole in either direction. (When the hole is large, a collective hydrodynamic flow occurs and the governing equation changes.) This process is described by the equation

$$\frac{dn}{dt} = -nvS + n_0 vS\,,$$

where S is the hole area, and v is the r.m.s. velocity of the molecules. Choosing the relevant variables, we again come to equation (13).

Equation (13) can be solved quite easily. For instance, for the initial condition $n = 0$ at $t = 0$ we obtain

$$n = n_\infty(1 - e^{-t/\tau})\,,$$

where $n_\infty = b/k$ and $\tau = k^{-1}$.

However, we failed to take into account that n, the number of molecules, is an integer. A popular Soviet poet Samuel Marshak has admonished children against such mistakes: "And the solution we heard was two men and two thirds".

This problem is essentially probabilistic. With the proper account of the integer value of n, considerations in terms of probability of changing the number of molecules in the vessel by one per unit time are necessary.

When the problem is probabilistic, the solution must be sought not as a smooth function $n(t)$ but rather as a probability distribution, i.e., a set $P(n, t)$ of the functions defined for integer values of $n = 0, 1, \ldots$ that do not depend smoothly on t.

Then the governing equation changes to

$$\frac{dP(n,t)}{dt} = bP(n-1,t) + k(n+1)\,P(n+1,t) - bP(n,t) - knP(n,t).$$

Indeed, the given number of molecules, n, can be produced either by adding one molecule to $(n - 1)$ molecules that are still in the vessel or by the loss of one molecule if their number was $(n + 1)$. The state of n molecules is changed when a molecule is either lost or gained. The probability of gaining a molecule does not depend on the number of molecules in a small vessel while the probability of a loss of a molecule is proportional to the number of molecules in the vessel. These presumptions lead to the given differential equation but we should assume that n is large; and, of course, neglect that n is an integer.

However, the classical equation for $n(t)$ can be derived from the chain of equations for $P(0, t), P(1, t), \ldots, P(n, t), \ldots$.

Consider an example where the solution can easily be given in a closed form: at the initial stages of filling a small vessel, the loss of molecules from it can be neglected. In this case

$$\frac{dP(n, t)}{dt} = A[P(n-1, t) - P(n, t)],$$

where A is the probability for a molecule to come into the vessel.

Initially, let $P(0, 0) = 1$ for $t = 0$, and $P(n, 0) = 0$ for $n > 0$, i.e., the vessel is initially empty. Then the exact solution reads

$$P(0, t) = e^{-At}, P(n, t) = \frac{1}{n!}(At)^n e^{-At}.$$

Let us check that the sum of all probabilities is unity. As follows from

$$1 + At + \frac{1}{2!}(At)^2 + \ldots + \frac{1}{n!}(At)^n = e^{At},$$

the sum is indeed unity:

$$\sum_{n=0}^{\infty} P(n, t) = e^{At}e^{-At} = 1.$$

The probability distribution obtained is known as the Poisson distribution.

Consider the behaviour of the solution of $At \gg 1$. The value of n for which P is maximal is determined by

$$\frac{\partial P(n, t)}{\partial n} = -\frac{P(n, t)}{n} + AtP(n, t) = 0,$$

which gives $n_{max} = At$, a natural and understandable result. This result cannot be considered rigorous since we have calculated the derivative as if $P(n, t)$ is a smooth function of n although n is valued as an integer. To be rigorous, one should calculate the average number of particles,

$$\bar{n} = \sum_{n=0}^{\infty} nP(n, t) = At .$$

A slightly less trivial operation is the calculation of the mean squared deviation,

$$\overline{(n - \bar{n})^2} = \sum_{n=0}^{\infty} (n - At)^2 P(n, t) .$$

This quantity turns out to be equal to $(At)^2$. For large $\bar{n}$, the distribution around $\bar{n}$ is very close to the Gaussian one. In particular, with the probability of 95% the value of n lies between $At + \mu(At)^{1/2} = \bar{n} + \mu\bar{n}^{1/2}$ and $At - \mu(At)^{1/2} = \bar{n} - \mu\bar{n}^{1/2}$ where μ is in the order of unity.

Similar results can be achieved, with more time and paper spent, in the problem when the motion of molecules is allowed in both directions through the hole. The fundamental difference is the fact that now the temporal growth of $\bar{n}$ saturates at some stationary value $\bar{n}_S$. The number of molecules approaches the stationary value according to the law

$$\bar{n} = \bar{n}_s(1 - e^{-\beta t})$$

where β = const.

However, just like the case of incoming particles, the distribution of the number of molecules around the average value is Gaussian. In particular, the number of molecules continues to fluctuate in time even when the average number $\bar{n}$ becomes stationary, $\beta t >> 1$. The final state is stationary in average but fluctuative. The plot of n versus t resembles a trajectory of a Brownian particle that wanders under the combined influence of elastic returning force and the random driving force.

2.8. SPATIAL DISTRIBUTION OF PARTICLES IN GASES

Many fields of physics, e.g., the theory of gases and solutions, the turbulence theory and cosmology often employ the notion of spatial homogeneity. In this section we analyze this notion in greater detail.

The air uniformly fills a room. Every equal volume contains an equal number of molecules, that is, 2.7×10^{19} molecules per cubic centimeter under normal conditions. It is obvious, however, that at microscopic scales, this homogeneity must be violated; the average number of molecules in an ultra-small volume 3×10^{-2} cm^3 is unity. In even smaller volumes, the numbers of molecules are smaller than unity, but a molecule cannot be divided between the two such volumes.

To the first approximation, in a volume with the average number of molecules equal to α, with $\alpha \ll 1$, resides one molecule with the probability α; the volume is empty with probability $1 - \alpha$. This situation is quite familiar as we have discussed such situations above. In particular, for an average number of molecules α (which can either be greater or smaller than unity but always positive) in the simplest case, we have found the probability distribution $P(0) = e^{-\alpha}$, $P(1) = \alpha e^{-\alpha}, \ldots,$ $P(n) = (\alpha^n/n!)\, e^{-\alpha}$. The coefficient of the exponential function for $P(n)$ is the nth term in the Taylor series for e^{α}. This probabilistic result can be checked experimentally by many consequent measurements of the number of molecules, i.e., by detection of many realizations of the same probabilistic situation. After that, the desired probability can be obtained as the fraction of realizations where the considered event occurs.

In the example with the number of molecules in a given volume, there is one more possible approach. Choose a small sub-volume and count the molecules within it; the result depends on time. Every time a molecule comes in or out of this sub-volume, a new realization occurs.

Thus, we can determine the function $n(t)$ of the total number of molecules versus time. This function is not smooth (Fig. 2.7) but consists of horizontal segments ($n = 0, 1, \ldots$) connected by vertical segments of unit length. The size of a molecule is considered small as compared with the size of the sub-volume. Therefore, the time of passage through the boundary of a sub-volume is small. In practically every case, except in very artificial situations, the average over many realizations and the

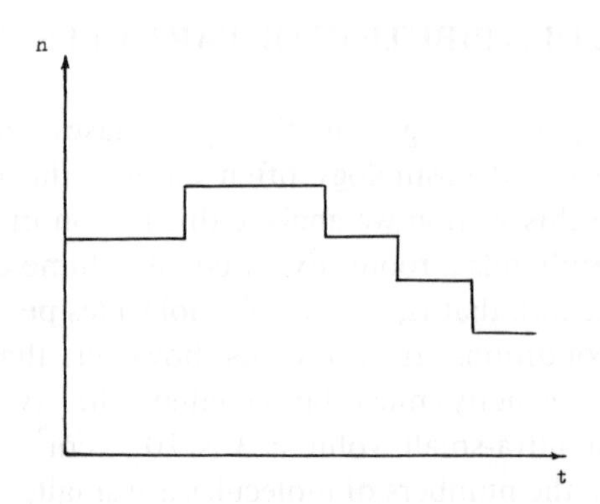

Fig. 2.7. The number of molecules within a given volume as a function of time.

time-average coincide. The scientific term for this intuitively clear assertion is "the ergodic hypothesis".

However, we spend time discussing the spatial distribution of molecules in a gas and not the formulation of the ergodic hypothesis. There are more interesting, principally new, questions in this spatial problem: How connected are the fluctuations in two neighbouring gas sub-volumes? How do these fluctuations depend on the size and shape of the sub-volume? How are they correlated in time?

If we apply these questions to a casino with many roulettes, the answers are simple and evident: the events (gain or loss) at every roulette are independent. If the probability of event A at one roulette is $P(A)$ and probability of event B at another one is $Q(B)$, the probability that both events occur at the respective roulettes, $R(A, B)$ is equal to the product, $R(A, B) = P(A)Q(B)$. However, the situation is different in the case of two neighbouring volume elements V_1 and V_2 chosen in a gas. There is considerable probability that a molecule that leaves the volume V_1 will enter the volume V_2 and vice versa. How does this affect the probability distribution for the respective numbers of molecules, n_1 and n_2?

In the analysis of the evolution of the number of molecules with time, the distribution over n is not the only important thing, i.e., the function of time when $n_1 \leqslant n \leqslant n_2$, but also the detailed dependence of n on t that is determined by specific processes that change n.

Consider a volume whose size is significantly greater than the time path of a molecule but still small when compared with the overall size of a vessel. Then one can distinguish the bulk motion of the gas and the acoustic waves in which density variations are accompanied by pressure variations. The characteristic time of the change of the enhanced density of the density below average is of the order of magnitude of the ratio of the considered volume size to the speed of sound. The waves which are shorter than the volume size are averaged out and do not contribute to the variations of the average density. The longer waves make such contributions but the number of independent long waves in a given vessel decreases with the wavelength. This explains why the main contribution comes from the waves whose lengths (the spatial period) are of the order of the volume size, but rarely from longer waves.

Another factor that is important is the fluctuations of the temperature (and entropy) in a given volume for constant pressure. A characteristic time of the temperature fluctuations and associated density fluctuations is of the order of the squared volume size divided by the diffusion coefficient.

These two types of fluctuations can be distinguished by observing light scattering in the gas because they produce different frequency shifts. The theory of this phenomenon was developed by Landau and Plachek in the thirties; its detailed discussion can be found in *Statistical Physics* by Landau and Lifshitz (1964).

Note also that when one considers not only density but also the averaged motion of molecules in a given volume, one should take into account not only the wave motion but also the transverse motion which is the vortex bulk motion of the gas. Only then can one arrive at the correct relation $m\bar{V}^2/2 = 3/2\, nkT$. In this respect an elementary volume of a gas or liquid behaves as an individual molecule. A similar result was discussed above where we considered Brownian motion of dust particles.

CHAPTER

THREE

DIFFUSION

3.1. THE DIFFUSION COEFFICIENT

Consider a solution of some substance, e.g., sugar in water. Denote the sugar concentration by g and its dimension by gcm^{-3}. The number of sugar molecules in 1 cubic centimeter is given by $n = Ag/\mu$, where A is Avogadro's number, 6×10^{23} mole^{-1}, and μ is the molecular weight; for sugar ($C_{12}H_{22}O_{11}$) $\mu = 12 \times 12 + 22 \times 1 + 11 \times 16 = 344$. In physical chemistry, the molar concentration is often used, $C = 1000\ g/A = 2.89\ g$ moles/litre.

Consider a weak solution in which the number of sugar molecules is small as compared with that of the water molecules everywhere in every part of the volume. The presence of sugar then weakly affects the concentration of water in the solution and the water concentration can always be considered constant. In addition, mutual interactions of sugar molecules can be neglected. Every sugar molecule interacts only with the neighbouring water molecules. The thermal motion of water molecules results in the chaotic motion of a sugar molecule, every direction of the motion being equally probable at any moment. The motion of a sugar molecule at any moment is practically independent of the motion at the preceding moment of time if the period between the considered moments exceeds the characteristic time (10^{-9} s) between the successive collisions of the sugar and water molecules.

As we have already noted, the motion of any sugar molecule is independent of the position and motion of the other sugar molecules.

It is remarkable that the analysis of the behaviour of an individual molecule turns out to be, in principle, more complicated than the analysis of the solution as a whole, i.e., simultaneous consideration of a tremendous number of sugar molecules, provided one is not interested in the fate of individual molecules.

The trajectory of an individual molecule will be discussed in Chapter 4. Following the historical development of the problem, here we discuss a macroscopic problem: considering the number of molecules in unit volume as a continuous variable, $n = n(x, y, z, t)$, we seek the law of evolution of n. A typical formulation of the problem assumes that an initial distribution $n(x, y, z, t)$ is prescribed for the moment t_0 and the problem is to find $n(x, y, z, t)$ for an arbitrary moment $t > t_0$.

The uniform time-independent concentration $n_0 =$ const. represents a solution. Indeed, the uniform solution is a state of thermodynamic equilibrium and therefore $n(x, y, z, t) = n_0$, independent of both coordinates and time, is a permissible distribution.

The next complex problem considers the distribution of n that initially, at $t = t_0$, depends only on x but not on y and z (one-dimensional problem).

Let us consider the surface $x =$ const. and count the number of sugar molecules that pass through this surface in both directions along y and z axes. Two different simplifications can be adopted in this analysis. The first model endows sugar molecules with fixed velovity v. The velocity vector can be directed at any point with equal probability. Thus, $\langle \mathbf{v} \rangle = 0$ and $\langle v^2 \rangle$ differs from zero. In the following the velocity $\tilde{v} = \langle v^2 \rangle^{1/2}$ is considered as a "characteristic" one.

The characteristic velocity of the motion along the x-axis is given by[a]

$$v_x = \sqrt{\langle v_x^2 \rangle} = \sqrt{\frac{\langle v^2 \rangle}{3}} = \frac{\tilde{v}}{\sqrt{3}}.$$

[a] The inelegant factor $3^{-1/2}$ arises due to the fact that the average velocity is defined, irrespective of the direction,

$$\langle v^2 \rangle = \langle v_x^2 \rangle + \langle v_y^2 \rangle + \langle v_z^2 \rangle .$$

This gives $\langle v_x^2 \rangle = \frac{1}{3} \langle v^2 \rangle$. For $\tilde{v} = \langle v^2 \rangle^{1/2}$ and $\tilde{v}_{x}, = \langle v_x^2 \rangle^{1/2}$ the equality in the text is recovered.

Let us now presume that there exists a certain average time period τ over which this velocity remains constant. In other words, we consider the molecules with a given value of v_x and this velocity is presumed to be fixed for a period τ. After this period, interaction with the chaotic motion of water molecules changes v_x and a new value v_x' can be arbitrary. It does not depend on the value of v_x during the preceding period τ. Of course, the parameter τ has a probabilistic character: some sugar molecules change their velocities after a period 0.5 τ, others after 1.5 τ. Of course, different molecules change their velocities to different extents. Approximately half the molecules have the velocity $v_x = 3^{-1/2}\tilde{v}$, the remaining ones have $v_x = -3^{-1/2}\tilde{v}$. During the period τ, they travel over the distance $\Delta x = \pm v_x \tau$. At a given moment, those molecules pass the surface $x = \text{const.} = x_0$ moving in the positive direction in which they started their free motion at a distance Δx from the surface, i.e., at $x_0 - \Delta x$. Consequently, the flux of molecules moving in the positive direction through the surface is

$$q_1 = v_x \frac{1}{2} n(x - \Delta x) .$$

The opposite flux is

$$q_2 = v_x \frac{1}{2} n(x + \Delta x) .$$

Thus, when n depends on x, the fluxes do not compensate each other! The net flux is given by

$$q_x = q_1 - q_2 = \frac{v_x}{2} [n(x - \Delta x) - n(x + \Delta x)] .$$

Expansion of $n(x)$ as a Taylor series leads to

$$q_x = -v_x \frac{\partial n}{\partial x} \Delta x = -v_x \frac{\partial n}{\partial x} v_x \tau = -v_x^2 \tau \frac{\partial n}{\partial x} = -\frac{1}{3} \langle v^2 \rangle \tau \frac{\partial n}{\partial x} .$$

The flux turns out to be proportional to the concentration gradient; the proportional coefficient is called the diffusion coefficient and denoted by D:

$$q_x = -D\frac{\partial n}{\partial x}.$$

The flux is directed toward the position with a smaller concentration. Thus, on average, the sugar molecules travel from the regions of greater n to the places where n is smaller.

This diffusion law established by Fick in 1855 can be easily generalized to the case of n dependent on all three coordinates,

$$\mathbf{q} = -D\nabla n. \tag{1}$$

Therefore,

$$q_x = -D\frac{\partial n}{\partial x}$$

is one of the components of the flux vector. The dimension of $\mathbf{q}$ is cm^{-2} s^{-1} while the dimension of n is cm^{-3}. Thus, D is measured in $cm^2\ s^{-1}$. The dimension of the diffusion coefficient does not depend on the specification of the flux and density, whether in weight, volume, or molar units, provided only these units correspond to each other ($cm^{-2}\ s^{-1}$ and cm^{-3}, or $gcm^{-2}\ s^{-1}$ and gcm^{-3}, or mole $cm^{-2}\ s^{-2}$ and mole cm^{-3}). From this dimension of D, we see that D can be constructed as $D \sim v^2\tau$ or $D \sim vl$, where v is the r.m.s. velocity [$cm\ s^{-1}$], τ is time between collisions [s], and l is the free path length [cm].

Let us now turn to an alternative schematic description (model) of motion of the sugar molecules. Consider the sugar solution as a sponge with fixed positions, "holes", where sugar molecules can rest. A molecule rests in this or that "hole" during an average period τ and then jumps into a neighbouring "hole" separated by the distance l. The jumps are instantaneous, i.e., they occur in a time τ', much smaller than τ. The flux is of course given by

$$\mathbf{q} = -D\nabla n, \quad D = \frac{l^2}{\tau},$$

where τ is a mean residence time in a "hole". It is convenient to introduce the probability $W\Delta t$ of leaving a "hole" during a short period Δt. The mean residence time is then expressed as $\tau = W^{-1}$. The dimension of W is s^{-1} and the diffusion coefficient can be written as

$$D = l^2 W.$$

Thus, we have considered two alternative descriptions of the motion:

(i) Free motion of sugar molecules with constant velocity during the period τ after which the velocity changes and the next period of free motion follows.

(ii) The rest of finite oscillations within a hole during the period τ and subsequent instantaneous jump and capture into another hole.

The former description is relevant for molecular diffusion in gases. The latter picture adequately describes the admixture motion in solids, e.g., a crystal. The diffusion of a substance dissolved in a liquid (sugar in water) is rather better described by the former picture. However, in both cases we consider the periods over which the probabilistic character of the considered processes is smoothened. As a result, in both cases the same diffusion law (1) emerges. The details of a molecular motion determine the particular value of D but the diffusion law (1) is always the same.

What is left is a trivial step: based on the flux $\mathbf{q}$, determine the evolution of the concentration. The conservation equation for the mass of the considered substance gives

$$\frac{\partial n}{\partial t} = -\nabla \cdot \mathbf{q}.$$

Therefore,

$$\frac{\partial n}{\partial t} = \nabla \cdot (D\nabla n). \tag{2}$$

For constant D,

$$\frac{\partial n}{\partial t} = D\nabla \cdot \nabla n = D\Delta n \,, \tag{3}$$

where Δ is the Laplace operator, or

$$\frac{\partial n}{\partial t} = D\left(\frac{\partial^2 n}{\partial x^2} + \frac{\partial^2 n}{\partial y^2} + \frac{\partial^2 n}{\partial z^2}\right).$$

Consider a simple particular form of $n(x)$ shown in Fig.3.1 and concentrate on three points, x and $x \pm \Delta x$. Introducing the Taylor expansions, we obtain a discrete representation of the Laplace operator (in one-dimensional case):

$$\frac{\partial n}{\partial t} = D\left(\frac{n_+ + n_-}{2} - n\right)\frac{1}{(\Delta x)^2}\,.$$

Now distinguish on the right-hand side the contributions from the molecules arriving and leaving at a given surface:

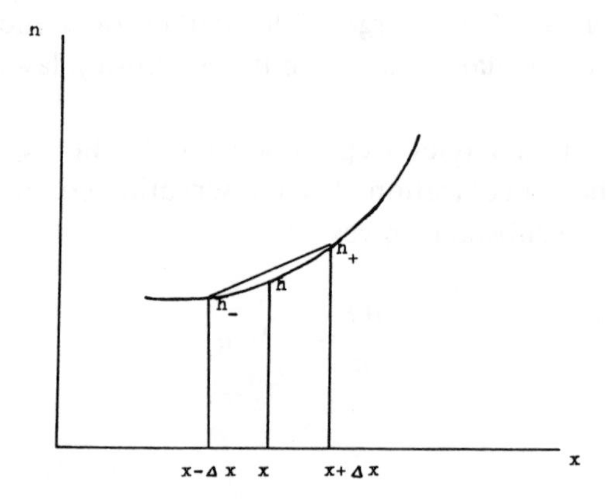

Fig. 3.1. Illustration of the discretization of the Laplace operator.

$$n(x, t + \Delta t) = n(t)\left[1 - \frac{D\Delta t}{(\Delta x)^2}\right] + \frac{n_+ + n_-}{2} \cdot \frac{D\Delta t}{(\Delta x)^2}.$$

This representation clearly indicates that diffusion — a random motion — levels the concentration.

In the early twentieth century, before the advent of computers, the following graphical method devised for the solution of the diffusion problem was proposed: choose Δt and Δx so that

$$n(x, t + \Delta t) = \frac{n_+ + n_-}{2}.$$

Each step is reduced to drawing a straight segment connecting the points n_+ and n_- with $n(x, t + \Delta t)$, representing the segment's middle point. This method can be generalized to three dimensional cases. Indeed, it is known that the Laplacian of a scalar function at a point $\mathbf{r}$ is equal to the difference (divided by r^2) between the function's value averaged over the spherical surface of radius r and its central value. Thus, diffusion both in one and three dimensions leads to $n(\mathbf{r}, t + \Delta t)$ becoming equal to the average value of $n(\mathbf{r}, t)$ over a certain sphere.

3.2. SOLUTIONS OF THE DIFFUSION EQUATION

In this section, we consider several typical situations and the corresponding solutions of the diffusion equation (3).

(i) A vessel with impenetrable walls. In this case, the normal component of the flux vanishes at the wall. Denote the vector normal to the wall by m. Thus, $q_m \equiv (\mathbf{q} \cdot \mathbf{m}) \equiv 0$ at the wall. It can be easily verified that in this case $N = \int n d^3\mathbf{x} = \text{const}$. A general solution with an arbitrary initial distribution, $n(x, y, z, t = 0)$, can be found by separation of the variables:

$$n(x, y, z, t) = \sum_k A_k n_k(x, y, z) e^{-\gamma_k t}$$

where the eigenfunctions n_k obey the following equation:

$$-\gamma_k n_k = \Delta n_k$$

with corresponding boundary conditions.

When n_k are restricted by the normalization condition $\int n_k^2 d^3\mathbf{x} = 1$, the coefficients A_k can be found from the following expansion of the initial distribution:

$$A_k = \int n(x, y, z, t = 0) n_k d^3\mathbf{x} .$$

The eigenfunction with the smallest eigenvalue by modulus, $\gamma = \gamma_0$, is constant:

$$\gamma_0 = 0, \quad n_0 = \bar{n} = \frac{N}{V} = \int n(x, y, z, t = 0) \frac{d^3\mathbf{x}}{V} .$$

For $t \to \infty$, any initial distribution tends to this eigenfunction at the rate determined by the value of γ_1. By the order of magnitude, $\gamma_1 = D/R^2$, where R is the vessel size.

(ii) Consider a vessel whose walls are membranes washed from the outside by solutions of a given concentration (Fig. 3.2). The concentration at one membrane is prescribed to be $n = n_1$ while $n = n_2$ at another membrane. The values of the derivatives $\partial n/\partial m$ at the membranes are not given in advance; they should be determined.

In this case separation of the variables leads again to a single steady solution:

$$\gamma = 0, \quad n = n(x, y, z), \quad \Delta n \equiv 0 .$$

However, in contrast to the case (i), now $n \neq$ const. It is only natural because the values of n are different on different membranes. In the simplest case of a cylinder with membranes at the ends (Fig. 3.3) we obtain $n = n_1 + (n_1 - n_2)\, x/L$, independently of y and z.

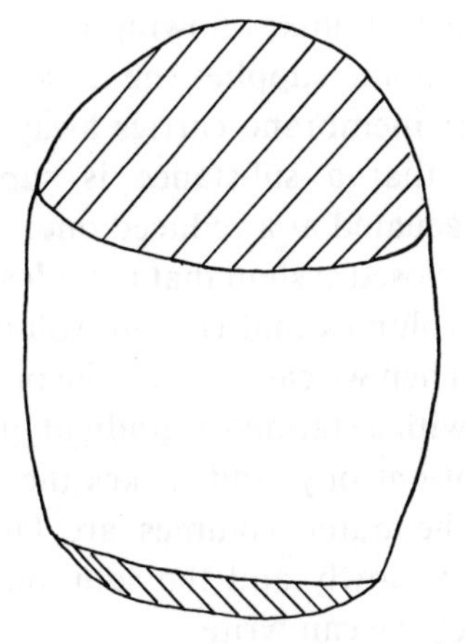

Fig. 3.2. A vessel with the ends covered with semi-permeable membranes (shaded).

The static solution is the equilibrium state and a stationary flux $Q\,[gs^{-1}]$ passes through the vessel in this state,

$$Q = \int (\mathbf{q} d\mathbf{s}) ,$$

where $d\mathbf{s}$ is the area element of the vessel's cross-section. The flux is the same for any cross-section within the vessel. At the vessel boundaries,

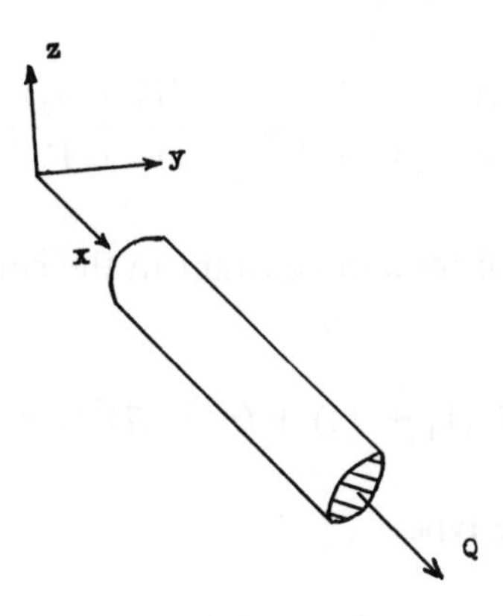

Fig. 3.3. A cylindrical vessel with membranes at the ends. The arrow indicates direction of the particle flux.

this state is maintained by fluxes flowing through the membranes: the flux through one membrane supplies the dissolved substance while the flux through the other membrane carries away the same amount. It is important, however, that a substance is supplied at an enhanced concentration and evacuated at a reduced one.

Consider the whole closed system that includes not only the vessel with membranes but also volumes that contain solutions washing the membranes from outside. Then we can see that the presence of the vessel with membrane walls and with a stationary gradient of concentration within it leads to the growth of entropy, and makes the whole system approach equilibrium. When the outer volumes are large, the concentrations within them vary only slowly and the solution for the inner vessel is nearly stationary. Then we can write

$$n_1 V_1 + n_2 V_2 = \text{const.}\,,$$

$$\frac{d}{dt}(n_1^2 V_1 + n_2^2 V_2) = D\int(\nabla n)^2 d^3\mathbf{x}\,, \tag{4}$$

where the integral on the right-hand side is taken over the inner vessel volume. These equations can be derived from the following considerations. The total amount of the dissolved substance is $N = n_1 V_1 + n_2 V_2$. Consequently, a mean concentration that is settled in equilibrium is given by

$$\bar{n} = \frac{N}{V} = \frac{N}{V_1 + V_2} = \frac{n_1 V_1 + n_2 V_2}{V_1 + V_2}\,.$$

Evidently, both N and $\bar{n}$ remain constant in the course of evolution. It can be seen easily that

$$n_1^2 V_1 + n_2^2 V_2 \equiv \bar{n}^2(V_1 + V_2) + (n_1 - \bar{n})^2 V_1 + (n_2 - \bar{n})^2 V_2\,.$$

The combination of the type

$$(n - \bar{n})^2 V$$

can be considered a convenient measure of deviation from equilibrium. The quantity $(n_1 - \bar{n})^2 V_1 + (n_2 - \bar{n})^2 V_2$ measures the deviation of the whole system from thermodynamic equilibrium. Taking everything into account the leading role is played by the large outer vessels whose volumes are V_1 and V_2. The amount of the dissolved substance kept in the smaller inner vessel can be neglected. The quantity $D(\nabla n)^2\, d^3\mathbf{x}$ characterizes the contribution of the processes occuring in the volume element d^3x to the total rate of evolution toward equlibrium. This quantity is analogous to the combination $\nu(\partial v_i/\partial x_k + \partial v_k/x_i)^2$ in hydrodynamics, with ν the viscosity and v_i the velocity components, that determines the rate of transfer of kinetic energy into thermal energy.

The equation that describes the approach to equilibrium under diffusion can also be written out easily for a single closed volume as

$$\frac{d}{dt}\int n^2 d^3\,\mathbf{x} = -2D\int (\nabla n)^2 d^3\mathbf{x}\,. \tag{4a}$$

The discussion of the role of mixing of diffusing liquid in a vessel is very instructive. Equations (4) remain applicable even when the liquid (solution) within the inner vessel is kept in motion. Obviously, motion of the liquid does not affect the value of $\int n^2 d^3\mathbf{x}$ in the absence of diffusion, as well as of any quantity of the type $\int f(n) d^3\mathbf{x}$. In the course of the motion, a small volume d^3x in which the concentration is n changes its position but neither n nor $d^3\mathbf{x}$ vary. (For simplicity, we consider an incompressible fluid.)

Consider a vessel in which the concentration varies in some regular way, e.g., as $n = a + bx$. Diffusion leads to a gradual equalization of the concentration. Asymptotically, for $t \to \infty$ we have $n = \text{const.} = \bar{n} = a + b\bar{x}$. Clearly, by mixing the liquid we can reach the equilibrium state earlier. Does this not contradict the assertion above that mixing cannot affect the value of $\int n^2 d^3x$? The answer is: No! The point is that mechanical mixing only indirectly affects the rate of evolution toward equilibrium in diffusive systems; it enhances gradients (cf. Fig. 3.3) and only then is diffusion intensified.

3.3. DIFFUSION IN INFINITE SPACE AND GREEN'S FUNCTION

The solution of the diffusion equation in infinite space can be found easily by separation of the variables. The eigenfunctions are plane waves $n_k(t = 0) = e^{ikx}$; particular solutions are given by $n_k = e^{ikx - Dk^2 t}$. The sequence of wavenumbers k forms a continuous spectrum which implies the necessity of employment of the Fourier integrals.

Consider a particular case when initially all the dissolved substance is concentrated at one point. In other words, n_0 is Dirac's delta function of position. One can consider the problem in one, two or three dimensions. In the real three-dimensional world, the one-dimensional distribution $n_0 = \delta(x)$ describes the initial concentration of the substance on a plane $x = 0$ (y and z are arbitrary on this plane) as shown in Fig. 3.4. Two-dimensional initial distribution $n_0 = \delta_2(s)$ arises when the substance concentrates at the z-axis. Finally, in three dimensions, the initial distribution of the substance can be confined to a point, e.g., the origin $x = y = z = 0$. In all the three cases, the coefficients of the Fourier integrals are constants. The Fourier integrals can be easily derived explicity. In one, two and three dimensions one obtains, respectively,

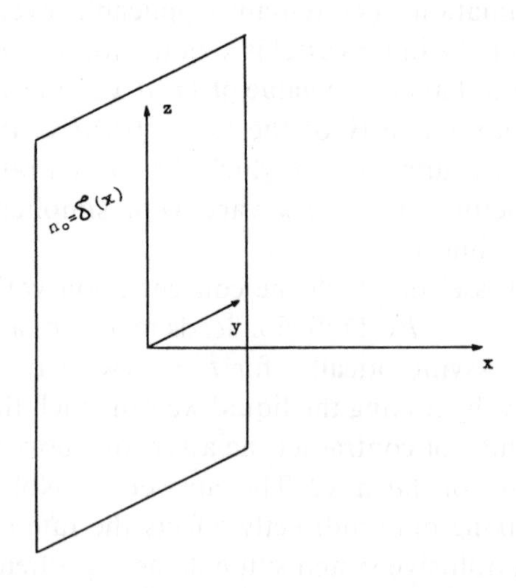

Fig. 3.4. The distribution $n(x, y, z) \sim \delta(x)$ describes the matter confined to the plane $x = 0$.

$$n_1 = \frac{1}{\sqrt{2\pi Dt}} \exp\left(-\frac{x^2}{2Dt}\right), \tag{5a}$$

$$n_2 = \frac{1}{2\pi Dt} \exp\left(-\frac{x^2 + y^2}{2Dt}\right), \tag{5b}$$

$$n_3 = \frac{1}{(2\pi Dt)^{3/2}} \exp\left(-\frac{x^2 + y^2 + z^2}{2Dt}\right). \tag{5c}$$

This solution is remarkable in many respects. Let us begin with the probabilistic interpretation of the diffusion process and this solution.

3.4. PROBABILISTIC INTERPRETATION OF DIFFUSION

Consider an individual molecule. At any given moment it is at some specific, exactly determined, position.

Following this particle, one can draw its trajectory. It can be seen in advance that this trajectory consists of many small straight line segments positioned at different angles. The length of an individual segment is equal to the crystal lattice period (of order several Angströms) when the particle diffuses in a solid or the free path length in a gas (in the order of 1000 Ångströms for a molecule of air under normal pressure). The common methods of calculating velocity, length and other quantities were developed for ordered smooth motions of cosmic bodies or of large Earth-bound objects and they can hardly be applied to such a curve as molecular trajectory.

We stress that apart from difficulties associated with unsmoothness of the curve, i.e., with the properties of an individual trajectory, there are difficulties of a more fundamental kind. They are associated with the irreproducibility of trajectories. Subsequently, let us set some particles free. Their trajectories are the same only statistically; every particle moves along its unique trajectory different from another. A direct and simple question, about where a given particle will be at the moment if it was initially at the origin ($t = 0$), has no definite answer when random forces participate in shaping its trajectory.

However a *probabilistic* answer can be given to this question.

Based on the statistical properties of the random force, one can find the probability for a particle to be here or there at the moment t. In the case of a crystal lattice, one can find the probability for a particle to occupy a certain lattice node; the nodes can be enumerated.

However, even when dealing with crystals, one never needs such accuracy. It is sufficient to determine the probability that a particle is within a certain volume $dx\,dy\,dz$. It is understood that dx, dy and dz are large when they are compared with the crystal lattice period, but small in comparison with the average distance travelled by the particles. Then we can introduce the probability density $P(x, y, z)$ so that the desired probability is

$$dW = P(x, y, z)\, dx dy dz\,.$$

Now, we should establish a connection between a statistical quantity, viz., the probability density, and the description of diffusion given above.

For the determination of the probability density, one can perform many an experiment by placing a particle at the origin and setting it free, simultaneously starting a stop-watch. When a particle moves, being driven by random kicks, one can record its position at a certain moment t. This experiment is repeated many, say N, times and then one divides by N the number of particles found within the volume element $dxdydz$. This gives the probability density (that obeys an obvious requirement $\int P d^3x = 1$).

Evidently, the motion of an individual particle does not depend on any other particle set free earlier or later. If so, why should we not save the experimental time?

Let us place all N particles at the origin and set them free simultaneously. Each particle moves along its own trajectory. At the moment t we record the positions of all particles. In this modification, the experiment is exactly the same as that used for the investigation of diffusion! The initial distribution is given by

$$n(x, y, z, 0) = N\delta(x, y, z)\,.$$

Since the diffusion equation is linear, its solution for N particles in three dimensions is as follows:

$$n(x, y, z, t) = N \frac{1}{(2\pi Dt)^{3/2}} e^{-r^2/2Dt},$$

where $r^2 = x^2 + y^2 + z^2$.

Hence, Green's functions of the diffusion equations (5a,b,c) are essentially probability densities of the associated statistical problem.

A very important result of this discussion is the fact that the mean squared path travelled in a given period is proportional to the length of this period. Therefore, the mean velocity,

$$\frac{\bar{r}}{t} = \frac{\sqrt{\bar{r}^2}}{t} \sim \frac{1}{\sqrt{t}},$$

decreases with time as the reciprocal square-root of time. Sometimes this random (diffusive, Brownian) motion is compared with the motion of a "perfect drunk": after each step, he forgets the direction to his home and takes the next step in an arbitrary direction. The total path R travelled in m steps can be represented as the vectorial sum of n vectors $r_1, r_2, \ldots, r_m$. Each vector r_i corresponds to one step, has the length r and a random direction.

Thus,

$$R = r_1 + r_2 + \ldots + r_m$$

$$R^2 = |R|^2 = (r_1 + r_2 + \ldots + r_m)^2$$

$$= mr^2 + r_1 \cdot r_2 + \ldots + r_i \cdot r_k + \ldots \quad (i \neq k).$$

For randomly chosen directions, all pairwise scalar products $r_i \cdot r_k$, $i \neq k$, vanish after averaging. This can be seen from the representation $r_i \cdot r_k = r^2 \cos \theta_{ik}$ while the average value of the cosine of a random angle is of course zero. Therefore,

$$\bar{R}^2 = mr^2.$$

Since each step has the same duration, m is proportional to time.

This simple discussion reveals a fundamental fact: the travelled path $(R^2)^{1/2}$ is proportional to the square-root of time:

$$\sqrt{\bar{R}^2} = r\sqrt{m} \sim \text{const.}\sqrt{t}\,.$$

This fact also follows from the solution of the diffusion equation. This result should not be applied carelessly. Evidently, it is inapplicable to periods shorter than the duration of a step because the statistical nature of this result requires that many events have occured. On the other hand, properties of the function exp $(-\ r^2/2Dt)$ imply formally that the superluminal motions are possible even if they have low probabilities. However, such a conclusion concerning hardly probable events cannot be made again due to the statistical nature of the diffusion law. The diffusion law is applicable only when the ratio $r^2/2Dt$ is not excessively large. Notice that for the motion $v = C$ this exponent becomes $(Ct)^2/2Dt = C^2t/2D$, and is very large even for moderate or small t.

The diffusion equation can be used for the solution of probabilistic problems in many more complicated cases.

One of the examples considers the random motion of particles near a wall to which they adhere after collision. Let the initial distance of a particle from the wall be r_0 and consider the probability for this particle to adhere to the wall in a given time. What distribution will the free particles have at the moment t? How do these probabilities behave for $t \to \infty$?

These problems are equivalent to the problem of diffusion with the boundary condition $n = 0$ at the wall and the initial condition $n_0 = \delta(r - r_0)$, where r_0 is the initial position of a particle.

A formal solution can easily be derived in the case of a flat infinite wall. The solution reduces to the introduction, apart from "positive" source at r_0, an additional "negative" source positioned symmetrically beyond the wall. When the wall coincides with the surface $x = 0$, the solution for $x > 0$ reads

$$n = \frac{1}{\sqrt{2\pi Dt}}\, e^{-\frac{y^2+z^2}{2Dt}} \left(e^{-\frac{(x-x_0)^2}{2Dt}} - e^{-\frac{(x-x_0)^2}{2Dt}} \right).$$

Obviously, this function satisfies the diffusion equation, reduces to $\delta(x - x_0, y, z)$ for $t = 0$ and identically vanishes on the wall $n(0, y, z, t) = 0$.

It can be shown that the total number of particles

$$N = \int_{\infty}^{0} dx \int_{-\infty}^{\infty} dy \int_{-\infty}^{\infty} dz\, n(x, y, z, t)\,.$$

tends to zero for $t \to \infty$ – all particles are bound to adhere to the wall! The solution given above allows us to follow the adhesion process and to find $N(t)$ with $N(0) = 1$ and $N(\infty) = 0$.

The result changes in the case of adhesion, for instance, for a spherical surface with particles initially positioned at some distances outside the sphere.

However, it is not our aim here to solve the various diffusion problems (and equivalently heat conduction problems).

The discussion above is aimed at demonstrating how the problems of the probability theory formulated for a single particle can be translated into the language of the diffusion theory.

3.5. DIFFUSION IN EXTERNAL FORCE FIELD

In this section, we discuss diffusion in the case where a particle's motion is affected not only by random forces but also by some deterministic ones. In the simplest cases, the latter can be considered time-independent and arising from a potential. The usual arguments start from the fact that under the action of any force, a particle acquires a certain acceleration. The motion is governed by Newton's equation and the sum of the particle's kinetic and potential energies is conserved. However, this habitual approach to describing motion under a given force is of little help in our problem.

In a fluid, an admixture molecule suffers irregular kicks from solvent molecules. As long as the molecule is at rest on average, the average numbers of kicks from different sides are equal to one another. However, when the molecule acquires some average velocity $\mathbf{u}$ under the action of an external force, the solvent molecules kick it more strongly, and more frequently in the direction opposite to **u** as compared with the accompanying direction. Thus, the force that arises counteracts the motion and is a resistance force. For an individual molecule and for a small velocity, this force is roughly proportional to the average velocity; the proportionality coefficient must be negative because the force direction is opposite to the velocity direction:

$$\mathbf{F}_{\text{res}} = -\alpha\mathbf{u}\,.$$

Therefore, quite soon after switching on the external force $\mathbf{F}_{\text{ext}}$, which remains constant, the particle accelerates to such a velocity that the total force (sum of the external force and the resistance) vanishes:

$$\mathbf{F}_{\text{tot}} = \mathbf{F}_{\text{ext}} + \mathbf{F}_{\text{res}} = \mathbf{F}_{\text{ext}} - \alpha\mathbf{u} = 0\,.$$

Thus, the average velocity of particles is given by $\mathbf{u} = \alpha^{-1}\mathbf{F}_{\text{ext}}$. The quantity α is called the resistance coefficient. The reciprocal, α^{-1} is called the particle mobility; it is equal to the ratio of the average velocity to the external force.

The equation for the particle trajectory is given by[b]

$$\mathbf{u} = \frac{d\mathbf{x}}{dt} = \frac{1}{\alpha}\mathbf{F}_{\text{ext}}(x)\,. \tag{6}$$

Considering a spatial particle distribution $n(\mathbf{x}, t)$, for the evaluation of the equation for n we can begin with the expression for the particle flux,

$$\mathbf{j} = n\mathbf{u}\,,$$

[b] Equation (6) is known in the history of physics as the basic equation of Aristotle's mechanics. The discussion above can therefore be interpreted, in the spirit of the correspondence principle, as the establishment of the range of applicability of pre-Newtonian mechanics. In a historical context, it is important to stress that actual differences between Newton's mechanics and the original approach of Aristotle are much deeper than the differences between equation (6) and Newton's second law. Basically, in the two millenia that elapsed between the eras of Aristotle and Newton, a new conceptual apparatus for mechanics was formulated.

and apply the conservation law for the particle number,

$$\frac{\partial n}{\partial t} = -\nabla \cdot \mathbf{j} = -\nabla \cdot (n\mathbf{u}) = -\frac{1}{\alpha} \nabla \cdot (n \mathbf{F}_{\text{ext}}(\mathbf{x})) . \tag{7}$$

This equation of partial derivatives is more complicated than the trajectory equation.

The trajectory is described as a characteristic of this equation in partial derivatives. Indeed, let the initial $t = 0$ and the particle distribution be $n(\mathbf{x}, t_0) = n_0(\mathbf{x})$. Consider a solution of the form

$$n(\mathbf{x}, t) = n_0(\mathbf{x}(t)) . \tag{8}$$

Substituting (8) in (7), we obtain for $\mathbf{x}(t)$ equation (6) with the initial condition $\mathbf{x}|_{t=0} = \mathbf{x}_0$.

What then is the advantage in replacing a simple equation for the trajectory of an individual particle with a more complicated equation for the particle distribution field?

In order to answer this question, we should have recourse to Section 3.4. The motion under random forces cannot be described by some specific trajectory. This description requires the introduction of the *probability* of the trajectory passing through a given point and the *probability density* at any given moment. This probability density obeys the diffusion equation, i.e., an equation in partial derivatives. Below we present the description of motion under the combined action of two forces, a deterministic force and random ones.

Since the presence of a random force necessitates the introduction of a partial-derivatives equation for $n(\mathbf{x}, t)$, it is clear that this is also necessary when the two forces are present. In preparation for this, we have introduced $n(\mathbf{x}, t)$ and the associated partial differential equation into a deterministic problem that could be solved in simpler terms, through a set of trajectories

$$\mathbf{x} = \mathbf{f}(\mathbf{x}_0, t) .$$

We see that the partial differential equation that describes the evolution of $n(\mathbf{x}, t)$ under the combined action of a deterministic external force and random (vanishing in average) forces has the form

$$\frac{\partial n}{\partial t} = -\nabla \cdot \frac{1}{\alpha} \mathbf{F} n + \nabla \cdot (D \nabla n) .$$

Consider the simplest one-dimensional case when, in addition, $\mathbf{F}$ is independent of time and has a potential, $\mathbf{F} = -\partial\Phi/\partial\mathbf{x}$ (this property is trivial in the one-dimensional case) and the coefficients $W = \alpha^{-1}$ (mobility) and D (diffusivity), both independent of time and position. Using the particle *flux* $\mathbf{j}$ as introduced above, we obtain

$$j_x = WFn - D\frac{\partial n}{\partial x} = -Wn\frac{\partial \Phi}{\partial x} - D\frac{\partial n}{\partial x},$$

$$\frac{\partial n}{\partial t} = -\frac{\partial j_x}{\partial x} .$$

Let us write out an important static solution: $j = 0$, $\partial n/\partial t = 0$. The variables separate giving

$$\frac{1}{n}\frac{\partial n}{\partial x} = \frac{\partial \ln n}{\partial x} = \frac{W}{D}\frac{\partial \Phi}{\partial x},$$

$$n = \text{const.} \exp\left(-\frac{W}{D}\Phi\right).$$

However, the particle distribution in stationary equilibrium must be

$$n = \text{const.} \exp\left(-\frac{\Phi}{kT}\right),$$

where k is Boltzmann's constant and T is the absolute temperature.

This brings us to Einstein's relation between mobility and the diffusivity:

$$D = kTW . \tag{9}$$

Let us check the dimensions: D has the dimension cm^2s^{-1}, kT is energy, gcm^2s^{-2}, and W is a ratio of velocity to force, $(cms^{-1}):(gcms^{-2}) = sg^{-1}$.

Boltzmann's equilibrium condition can be extended to the three-dimensional case:

$$\mathbf{j} = WFn - D\nabla n = -Wn\nabla\Phi - D\nabla n\,. \tag{10}$$

Condition $\mathbf{j} = 0$ leads to $n = \text{const.}\exp(-W\Phi/D)$. The solution with $\mathbf{j} = 0$ agrees with the boundary condition at fixed impenetrable boundaries, $\mathbf{j}_m = 0$.

Einstein's relation $D = kTW$ has played an important role in the development of the molecular kinetic theory. The mobility of a sphere in water can be determined with the help of the hydrodynamic theory of motion of a sphere in a viscous fluid. Stokes' formula gives $\alpha = 6\pi R^2\eta$, where R is the radius of the sphere and η is the fluid viscosity. The diffusion coefficient can also be estimated experimentally. Using the results of suitable experiments, Einstein considered sugar dissolved in water and used the measurements of the solution viscosity as a function of its concentration in order to refine the estimate of R. As a result, Avogadro's number was obtained: the number of molecules in one gram-mole was 6×10^{23}. Although the accuracy was not very high, this result shows that the molecular theory is applicable not only to gases but also to liquids.

New papers of Einstein and other authors were published shortly after in which ultra-small spheres in water were considered. The motion of small particles which are, nevertheless, millions of times larger than individual molecules, was first investigated by an English botanist Brown (see Chapter 4). He had even considered the hypothesis that forces specific to living matter play a role in this phenomenon. However, in the early 20th century, it was shown that he had actually investigated a manifestation of thermal motion.

J. Perrin's experiments with Brownian particles (whose size was exactly known) also revealed their experimental distribution over height in a gravity field which was consistent with the Boltzmann distribution. This allowed him to measure the product kT. The pressure and the volume of an ideal gas are related by $PV = RT$; the ratio R/k gives Avogadro's number.

Let us return to the diffusion equation in this force field. The following relations can be used to define the stationary state: $\partial n/\partial t = 0$, $n = n(\mathbf{x})$. Equation (7) shows that these relations are satisfied when $\nabla \cdot \mathbf{j} = 0$; the more stringent constraint $\mathbf{j} = 0$ is excessive. However, the solution with $\mathbf{j} \neq 0$ essentially differs from what was obtained above.

Indeed, when the force is a gradient of some potential Φ, we can write $\mathbf{j} = -n[\nabla\Phi + D\nabla \ln n]$. The vector $\mathbf{j}/n$ is potential.[c] In a closed volume, such a vector can differ from zero only when j_m is non-vanishing at the volume boundaries.

The physical meaning of this requirement is the following: the *static* state, that is, the thermodynamic equilibrium state, can be maintained infinitely long in a closed volume without external energy sources. Meanwhile, the *stationary* state ($\partial/\partial t = 0$, $\mathbf{j} \neq 0$) is possible only as long as external sources of the total energy or free energy are available.

As an example, consider two connected large vessels with solutions of distinct concentrations n_1 and n_2, and in this case a potential field Φ. In the stationary state, the concentration in one vessel continuously decreases while that in the other vessel increases. Therefore, entropy grows and free energy of the whole system decreases.

Consider now the case of non-potential force, $\nabla \times \mathbf{F} \neq 0$. Then a stationary flow is driven within a closed vessel ($j_m = 0$ at the walls). This stationary flow dissipates the energy supplied by the source of the vortex force. Thus, consider the diffusion in the external potential force supplemented by fluid motion.

The velocity of motion enters the diffusion equation through $j_m = nu$. The flow is usually (at least in liquids) incompressible, $\nabla \cdot \mathbf{u} = 0$. Therefore, for $n =$ const. we have

$$\frac{\partial n}{\partial t} = -\nabla \cdot \mathbf{j} = -n\nabla \cdot \mathbf{F} - \nabla \cdot (D\nabla n) \equiv 0\,.$$

The equilibrium with n independent of both position and time is not violated by incompressible fluid motion.

[c] We say that a vector is potential when not only $\nabla x \mathbf{j} = 0$ at any point within a vessel but also $\oint \mathbf{j} d\mathbf{l} = 0$ along any closed contour. When the vessel has a complicated topology, the second definition is generally used.

However, as we have already noted, if the admixture distribution is non-uniform at $t = 0$, then even the incompressible motion accelerates evolution towards the equilibrium. If the fluid was at rest, the Boltzmann distribution would arise for $t \to \infty$. In this case even a stationary incompressible flow affects $r(x)$ for $t \to \infty$. What arises is the diffusive dissipation; the fluid motion serves as an energy source.

3.6. OSMOTIC PRESSURE

The form of the diffusion equation with inclusion of external forces suggests that the diffusion itself can be interpreted as a motion under kinetic pressure.

Indeed, the diffusive flux is given by

$$\mathbf{j} = -D\nabla n.$$

Recall the Einstein relation $D = WkT$ and recast the diffusive flux as

$$\mathbf{j} = n\mathbf{u}D = n\left(-\frac{W}{n}\nabla kTn\right).$$

The combination kTn is nothing else but the pressure of an ideal gas of temperature T and concentration (particle number density) n.

The pressure produced by a substance dissolved in a liquid is usually called the osmotic pressure.

Thus, consider a dissolved substance that exerts the pressure P only on itself. Then the gradient of this pressure taken with an opposite sign, $-\nabla P$, presents the force per unit volume acting on this substance.

When the concentration is n, the quantity $-1/n\ \nabla P$ is the force per single particle of the dissolved substance:

$$\mathbf{F}_1 = -\frac{1}{n}\nabla P. \tag{11}$$

According to the definition of mobility, a particle under this force acquires the velocity $W\mathbf{F}_1$. Substituting W expressed through D, we obtain the particle velocity u. Then relations

$$\mathbf{j}_D = n\mathbf{u}; \quad \frac{\partial u}{\partial t} = -\nabla\cdot\mathbf{j}_D = D\Delta n\,,$$

lead to the diffusion equation.

Thus, we have presented a mathematically correct procedure of derivation of the diffusion equation based on the concept of osmotic pressure. We would like the reader to feel how strange this derivation is! Indeed, we have begun our discussion of the diffusion theory with the presumption that the particles of the dissolved substance do not interact with one another. Now we introduce the *pressure* determined by the particle number density and the force determined by the gradient of the pressure which acts on a given particle!

This approach is intrinsically contradictory but it cannot be ignored since the result is correct. Notice that the *pressure* by itself is not necessarily associated with interactions and collisions of molecules. This can be seen already from the form of the expression of the ideal gas pressure, $P = nkT$, which does not include the collisional cross-section.

For arbitrary velocity distribution the stress tensor of an ideal gas

$$T_{ik} = \sum_n m^{(n)} u_i^{(n)} u_k^{(n)}\,, \tag{12}$$

depends on particle velocities (momenta) rather than on their interactions. Of course, an interaction of some kind is required in order to transfer the pressure or a component of the stress tensor to a rigid wall. The molecules must experience a repulsion force and rebound when they hit a wall.

However, we emphasize again that the pressure inside a gas volume is not associated with mutual interactions of the gas molecules. This argument is in favour of the description of diffusion as a motion driven by gradients of osmotic pressure.

But there is one more objection to the association between diffusion and osmotic pressure. Employing the osmotic pressure concept, one would say that all particles slowly move along some specific trajectories.

If, for instance, all particles were initially concentrated at the origin at $t = 0$, then their trajectories correspond to the overall expansion

$$x = A\sqrt{t}, \quad y = B\sqrt{t}, \quad z = c\sqrt{t}$$

(Fig. 3.5). Every particle is characterized by its own set of coefficients A, B and C.

Is this picture adequate? The macroscopic density distribution that follows from this motion law exactly coincides with both those calculated from the diffusion equation and observed distributions. This can be verified by evaluating the particle distribution over the coefficients A, B, and C at some moment t after the beginning of the motion (it is given by $n(t_0) \propto \exp(-r^2/2Dt_0))$ with $r^2 = (A^2 + B^2 + C^2)t$ with subsequent determination of particle positions and their number density at an arbitrary moment. The result predicts that at the points, whose distances from the origin are proportional to $\sqrt{t}$, i.e., equal to $r_0(t/t_0)^{1/2}$ with different values of r_0, the density decreases as $t^{-3/2}$ or inversely proportional to the cube of the travelled distance.

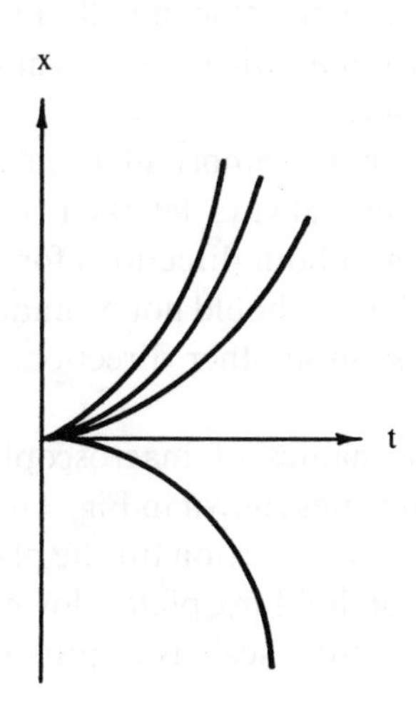

Fig. 3.5. Trajectories of particles according to the theory of osmotic pressure.

If, however, we enumerate particles or mark them by different colours and follow their motion individually, we would discover that they move in an entirely different fashion. The particles which reside within a certain small volume $dxdydz$ initially at $t = t_0$, (or, equivalently, have a certain set of values of A, B and C) perform random walks. At any later moment t, they do not travel to the corresponding small volume situated $(t/t_0)^{1/2}$ times farther from the origin. The particles travel to other positions in the course of their chaotic motion along trajectories that are far from being straight lines emerging from the origin. Any given volume is filled with particles that originate from the volumes whose positions are not related by the $t^{1/2}$ law. Nevertheless, the density mysteriously changes in exactly the same manner as predicted by the theory of smooth motion under the action of osmotic pressure without allowance for chaotic molecular collisions.[d]

In the theory of solutions, the osmotic pressure reveals another aspect. Imagine a semipermeable membrane, i.e., a film that allows one substance to pass through it (the solvent in our case, e.g., water) and forbids the penetration of another substance, e.g., dissolved sugar or salt molecules.

Such membranes are by no means exotic novel high-technology products. Various semipermeable membranes are common in living organisms. Life, at least in multi-cellular organisms, would be impossible without these membranes.

The membranes have an important property: if they selectively let some substances in, they always let them out! Semipermeable membranes are always open in both directions for a selected substance.

Thus, semipermeableness should not be understood as openness in one direction and closeness in another direction, with respect to the same substance.

Considering the mechanics of macroscopic bodies, one can easily conceive a trough with plates shown in Fig. 3.6, devised in such a manner that the balls rolling in one direction lift the plates, while the balls rolling in the opposite direction hold the plates down and rebound.

Such a device of molecular scale is in principle impossible!

[d] Allowance for molecular collisions *together* with osmotic pressure would amount to accounting *twice* for the same effect!

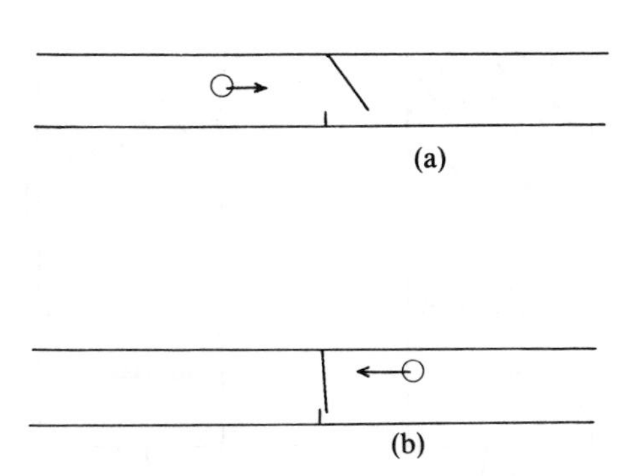

Fig. 3.6. The construction of a trough which allows a ball to move to the right (a) but not to the left (b).

If a membrane with one-way leak were possible, a gas could flow from a vessel under lower pressure to a vessel under higher pressure. A *perpetuum mobile* would then beautify our life. This would be the *perpetuum mobile* of the second kind: when expending work on rotating a turbine the gas would be cooled. Such engine would require a thermal energy supply and the energy conservation law would not be violated. However, such a device would violate the second law of thermodynamics, the law of entropy growth.

Let us, however, return to a real physical experiment. The phenomenon of osmosis can be observed in an apparatus shown in Fig. 3.7 where a semipermeable membrane separates the vessel to the left filled with a pure solvent from the vessel to the right filled with a solution. For any initial levels of liquids in both vessels, liquid penetration through the membrane leads to an equilibrium configuration.

In this equilibrium state, the liquid level is higher in the vessel that contains a solution.

The pressure exerted on the membrane from the right is higher than that from the left by the value of the *osmotic pressure* $P_0 = nkT$, where n is the concentration of the dissolved substance measured in molecules per unit volume.

When the concentration C is measured in moles per litre and the temperature is 0° centigrade (273 K), the osmotic pressure is equal to $P_0 = 22.4C$ atmospheres. The quantity, introduced simply for a more

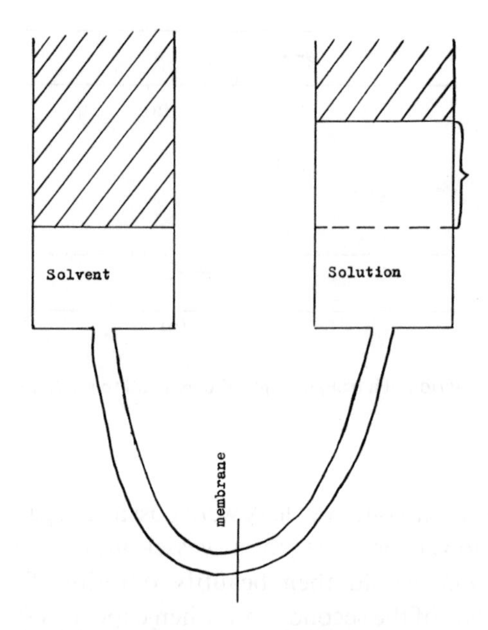

Fig. 3.7. The measurement of osmotic pressure.

convenient description, materializes and can be measured directly with a manometer or even with a rule from known liquid levels, densities and the gravity acceleration. Fig. 3.8 illustrates how osmosis helps a plant to absorb solutions from the soil by lifting them up.

The theory of osmosis commemorates many glorious achievements of a physical chemistry. One of the classical results is that osmotic pressure of a solution of the common salt, NaCl, is twice as large as predicted based on its molecular weight, $23 + 35.5 = 58.5$. This result[e] was interpreted as an indication that in the solution every molecule of NaCl dissociates into two ions, Na^+ and Cl^-, each of which is surrounded by several water molecules and moves independently; for this reason the pressure is enhanced.

[e] We hope that the reader would not be surprised by the fractional molecular weight of chlorine: it has two equally abundant isotopes.

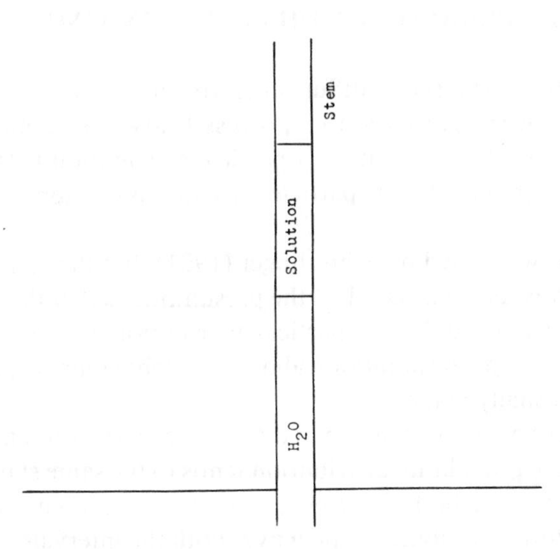

Fig. 3.8. The osmotic pressure allows a plant to draw liquid from the soil.

The very notion of osmosis stems from a Greek word meaning "to push". Molecules, atoms or ions that cannot penetrate a membrane fiercely push it, thus producing an enhanced pressure.

This effect can be easily understood in the case of a gas mixture. If gas A cannot penetrate a membrane and the vessel beyond the membrane is evacuated, it is clear that the membrane experiences a pressure P_A from one side and zero pressure from another.

Let us add some gas *B* which can penetrate the membrane. Naturally, in an equilibrium state the concentration and pressure of the gas *B* will be the same on both sides of the membrane and only the molecules of *B* passing in different directions coincide with each other.

The presence of *A* molecules on one side of the membrane has no effect on this process. Therefore, the pressure on one side of the membrane becomes P_B and that on the other side is $P_B + P_A$. The pressure difference remains as P_A. This is the osmotic gas pressure. It is important to note that mutual interactions of molecules are inessential in gases.

It is remarkable that a simple expression for osmotic pressure remains valid for solutions where a dissolved substance strongly interacts with a solvent.

3.7. ON REVERSIBILITY OF DIFFUSION IN TIME

In a thermodynamical sense, diffusion is irreversible in time: for a large number of diffusing particles, this process leads to smoothing, with a 100% chance of inhomogeneities of particle distribution $n(t)$ occurring. At any earlier moment, the particle distribution inhomogeneity was sharper.

However, it was noted by Schrödinger (1931) that the conclusion that diffusion is irreversible is based on the presumptions that the initial state of particles is fixed and that the particle distribution tends to a stationary one for $t \to \infty$. No presumptions about the behaviour of particles for $t \to -\infty$ are usually made.

Schrödinger has shown that when diffusion occurs over a finite spatial interval and the probability distribution tends to the same stationary and ergotic distribution, both for $t \to \infty$ and $t \to -\infty$ (Schrödinger considered a uniform distribution), then over both the intervals $-\infty < t < t_0$ and $t_0 < t < \infty$, the probability distributions evolve identically, i.e., diffusion turns out to be reversible in time.

This question was analyzed in detail by A. Kolmogorov (see Kolmogorov, 1986) and Yaglom (1949) for a general case of multidimensional Brownian motion.

Let there be given the stationary probability distribution density $p(x)$ of a diffusion process and the probability $f(t, x, y)dy$ of changing, during the period t, the position from x to some point from the interval $(y, y + dy)$ such that $f(t, x, y) \to p(\mathbf{x})$ for $t \to \infty$. Knowing the position at the moment we can determine the conditional probability density for the initial distribution of position x, $h(t, x, y)$. This function is uniquely determined by the equality

$$h(t, x, y)\, p(y) = p(x)\, f(t, x, y)\,.$$

In these terms, the problem of time-reversibility of diffusion is formulated as the problem of finding the conditions under which the equality $h(t, x, y) = f(t, x, y)$ holds, if the function $f(t, x, y)$ obeys both the diffusion equation (the Fokker-Planck equation)

$$\frac{\partial f}{\partial t} = v(x)\frac{\partial f}{\partial x} + D(x)\frac{\partial^2 f}{\partial x^2} \qquad (13)$$

and the adjoint equation

$$\frac{\partial f}{\partial t} = -\frac{\partial(v(y)f)}{\partial y} + \frac{\partial^2}{\partial y^2}(D(y)f)\,. \tag{14}$$

The stationary distribution $p(y)$ should obey the equation

$$-\frac{\partial(vp)}{\partial y} + \frac{\partial^2}{\partial y^2}(Dp) = 0 \tag{15}$$

which follows from (14), and the obvious condition $\int p(y)dy = 1$.

Based on these relations, Kolmogorov proved that for the reversibility of diffusion it is necessary and sufficient that the condition

$$v(x) = D^{-1/2}\frac{\partial}{\partial x}\ln\left(D^{-1/2}p(x)\right)$$

is fulfilled.

Above, we have discussed molecular diffusion. The concept of diffusion is also widely used in the problems of transfer of scalar or vectorial quantities, e.g., admixtures, temperature, or magnetic field, by hydrodynamic turbulent flows (see Chapter 7). Like the effects of the chaotic molecular motion, the effect of the random small-scale flows is approximately described as diffusion. Zeldovich (1982) discusses the symmetry properties of the turbulent diffusivity under time reversal. In the framework of a simple model of scalar transfer, the problem is solved by expressing the turbulent diffusion coefficient D_T in terms of the molecular diffusivity D:

$$D_T = D\left(C_1\frac{l^2v^2}{D^2} + C_2\right)^{1/2},$$

where v is the root-mean-square velocity, l is the correlation scale of the turbulence and C_1 and C_2 are constants of order unity. In the limit $D \ll lv$ the commonly used expression $D_T = C_1 lv$ is recovered. Thus, the turbulent diffusion coefficient has exactly the same symmetry properties as the molecular diffusivity.

CHAPTER

FOUR

THE BROWNIAN MOTION

The Brownian motion often serves as a paradigm of randomness. Unexpected turns of a particle path indicate the unpredictable character of the acting forces. It can be seen that under the influence of random kicks in a uniform isotropic medium, the Brownian particle would mark time within some finite vicinity of its initial position. Actually, the trajectories of a Brownian particle are very tangled and the mean value of its coordinate is zero. But in the sense of the mean square, the particle moves away from its initial position: the radius of the above-mentioned vicinity grows with time. Studies of the Brownian motion have a very interesting history commencing with a paper published by the botanist Robert Brown FRS (1773–1858) in 1828 (Brown, 1828). Using a microscope furnished with a lens of focal distance about 1 mm, Brown discovered a continuous disordered motion of minuscule pollen particles. Being a devoted botanist, he checked if this motion is an expression of life. Having tested glass and mineral particles, Brown rejected a vitalistic origin of the motion but decided that he has discovered the "active molecules". In reality, however, Brownian particles are subject to kicks from a great number of molecules, and their behaviour is simply a play of chance.

The Brownian motion is deeply investigated from the physical and mathematical points of view as an example of stochastic processes.

In 1900, L. Bachelier (Bachelier, 1900) pointed out that the probability distribution of a particle in Brownian motion is described by the diffusion law. Physically transparent formulation and explanation of this fact were given in 1905 and 1906 by A. Einstein, who proceeded from the idea that the Brownian particle moves due to random, uncorrelated kicks from many surrounding molecules. Independently and in simpler ways, these concepts were developed by M. Smoluchowsky (1906).

The kind of motion discovered by Brown has found its place in many fields of science, acquiring there various exotic forms which were very different from the patterns observed by the English botanist. The use of statistical distribution of Brownian particles to measure the integration in the infinite-dimensional space has provided a basis for numerous applications. The penetration of the Brownian motion is attributed to P. Levi and N. Wiener. An idealized random trajectory of an individual weightless particle is called the Wiener path. It has surprising properties – being continuous and non-differentiable at every point. R. Feynman has applied the functional approach to quantum mechanics and M. Katz has done this in statistical physics. Brownian trajectories can be encountered, often in a concealed form, in modern statistical radiophysics, turbulence theory and magnetohydrodynamics.

The works of Einstein and Smoluchowsky were aimed at substantiating the molecular kinetic theory as an example of analysis in the Brownian particle motion. When molecules bombard a sufficiently large body, its averaged momentum is close to zero and the body is practically at rest. The surface area of a Brownian particle is so small that random kicks of molecules are not balanced and, in contrast to macroscopic bodies, the particle moves. With the use of simple results of classical diffusion theory, certain basic molecular characteristics can be deduced by observing their motion. Chapter 2 is devoted to the classical theory of diffusion. Our main goal in this chapter is to discuss, after having recalled Einstein's description of the Brownian motion, more recent and more complicated approaches to this motion, especially the Wiener paths and the associated measure. Until now, many remarkable and constructive results derived in this field by mathematicians have not penetrated the daily practice of physicists. Our desire to build a bridge warrants our simplified discussion which may embarass a high-brow mathematician. Numerous applications of the concepts developed in this chapter are found in other parts of the book in forms that are rather unexpected, e.g., the connection between the Brownian motion and a ball of lightning (Chapter 5) or the role of the Wiener paths in the hydromagnetic dynamo (Chapter 9).

4.1. EINSTEIN'S DESCRIPTION OF BROWNIAN MOTION

Let us denote the number density of pollen particles at the point x and a time moment t as $n(t, x)$[a] and the fraction of particles which shift from point x to $x + \Delta$ in a short period τ as $\varphi(\tau, \Delta)$. Then we have at the moment $t + \tau$

$$n(t + \tau, x) = \int_{-\infty}^{\infty} n(t, x - \Delta)\, \varphi(\tau, \Delta)\, d\Delta\,. \tag{1}$$

Particle conservation dictates that $\int_{-\infty}^{\infty} \varphi d\Delta = 1$. It is natural to expect that the function $\varphi(\Delta)$ is symmetric, i.e., $\int \varphi \Delta d\Delta = 0$, and the probabilities of large shifts are small, i.e., $\varphi(\Delta)$, and concentrated near the small values of Δ. This allows us to expand the density in the Taylor series:

$$n(t, x - \Delta) = n(t, x) - \frac{\partial n}{\partial x}\Delta + \frac{\partial^2 n}{\partial x^2}\frac{\Delta^2}{2} - \ldots.$$

After substituting this expansion into (1) and performing the integration, we transfer the first term to the left-hand side and keep only the first non-vanishing term on the right-hand side. Due to the smallness of τ, we obtain

$$\frac{\partial n(t, x)}{\partial t} = D\frac{\partial^2 n}{\partial x^2}, \tag{2}$$

where we have introduced the notation $D = (2\tau)^{-1} \int \varphi \Delta^2 d\Delta$ (the diffusivity). Thus, the number density of pollen particles obeys the diffusion

[a] The problem can be formulated in another way: considering a single particle, $n(x, t)$ can be interpreted as the probability of the particle residing within the interval $(x, x + dx)$ at the moment t. When mutual interactions of particles can be neglected, as is implicity presumed in Einstein's formulation, both formulations are equivalent.

equation. Following from (2), if initially the particles were concentrated at the point x_0, then at the moment t we have

$$n(t, x) = \frac{1}{(2\pi Dt)^{1/2}} \exp\left\{ -\frac{(x - x_0)^2}{2Dt} \right\}. \tag{3}$$

This implies that the mean square of deviation of the Brownian path is given by $\langle (x - x_0)^2 \rangle^{1/2} = (Dt)^{1/2}$.

It is interesting and instructive that the latter result follows from a simple phenomenological argument (see, e.g., Gardiner, 1985). Consider a particle of mass m and radius R whose motion is driven by disordered kicks of surrounding molecules. Following Langevin, such motion can be described by introducing the random force $\mathbf{F}$ uncorrelated to the particle position: $\langle \mathbf{xF} \rangle = 0$. Newton's law yields

$$m\frac{d^2x}{dt^2} + 6\pi\eta R\frac{d\mathbf{x}}{dt} = \mathbf{F},$$

where the Stokes friction force has been taken into account (η is the friction coefficient of particles against the medium). Multiplying this equation by x, averaging, and taking into account that $\langle m\mathbf{x}^2/2 \rangle = NkT/2$, where N is Avogadro's number, we obtain for large t

$$\langle (\mathbf{x} - \mathbf{x}_0)^2 \rangle^{1/2} \simeq \left(\frac{NkT}{3\pi\eta R} t \right)^{1/2} = (Dt)^{1/2}, \tag{4}$$

where $D = NkT/3\pi\eta R$. Using the results of refined experiments, J. Perrin (1924) has verified this relation (and similar ones) and confirmed the value of Avogadro's number $N = 6 \times 10^{23}$ which plays an exceptionally important role in molecular physics and chemistry.

Notice that this seemingly simple "physical" reasoning is based on implicit audacious presumptions. In particular, simultaneous employment of both the viscosity and random force concepts seems to be unjustified. Friction in gases itself arises as a result of random forces. In a rarefied gas (Knudsen's range of parameters), the Stokes formula $6\pi\eta R(dx/dt)$ is inapplicable, while in a dense gas or in liquid molecular kicks, it cannot be considered independent.

In contrast to this derivation, the diffusion approach given above is based on solid ground.

Thus, the Brownian motion can be considered a random process whose probability distribution is described by the diffusive Gaussian law. This result clarifies both Einstein's approach which introduces the particle number density and derives relations (1)–(3) and the expression of the mean square deviation (4). However, there are possibilities to proceed further and consider individual realisations (trajectories) of this random process.

4.2 THE WIENER PATH

Let us denote $w_t(\omega)$ as a random coordinate of a Brownian particle, where t stands for time and the parameter ω specifies a particular realization of the random process. For a given moment $t = t_*$, the coordinate $w_{t_*}(\omega)$ is a random quantity. For a given ω, w_t is a deterministic function of t which is called the Wiener path.[b] This tribute to N. Wiener explains why the coordinate is denoted so unusually, w_t, instead of the more common $x(t)$.

A random Wiener process is defined by the following properties: $w_0(\omega) = 0$, $w_t(\omega)$ is the Gaussian process such that $w_{t+\tau}(\omega) - w_t(\omega)$ has zero mean value and a variance τ (see, e.g., a modern mathematical discussion in Hida, 1980).

Thus, the mean value of the coordinate along a Wiener path vanishes while its mean square grows proportionally to time. Since for a regular motion, the mean square is proportional to the squared time, this indicates that a path is highly irregular.

A Wiener path, being a trajectory of diffusion process, is continuous. However, it is non-differentiable at any moment. Indeed, the probability that $|w_{t+\Delta t} - w_t| > C\Delta t$ is given by the integral $2\int_{C\Delta t}^{\infty}(2\pi\Delta t)^{-1/2}$

[b] To be pedantic, the path is, of course, a line described by the function w_t.

$\exp(-w^2/2\Delta t)dw$ which tends to unity for $\Delta t \to 0$. Thus, for any constant C, the derivative dw_t/dt is almost certainly infinite at any moment t. This result can be reached in the following less rigorous manner. A typical value of the coordinate w_t is roughly proportional to $t^{1/2}$. Therefore, $\Delta w \sim (\Delta t)^{1/2}$ for infinitesimal increments,[c] rather than $\Delta w \propto \Delta t$, as required for the existence of common derivative. It is said that a Wiener path has the derivative of the order 1/2. (For comparison, note that in the Kolmogorovian turbulence for infinitely weak viscosity, the velocity is proportional to $l^{1/3}$, i.e., it has the spatial derivative of the order 1/3.)

Saying that a path is typical, we imply that the exponential function in the Gaussian probability distribution for w_t significantly differs from zero, for $|w_t| \leqslant Ct^{1/2}$. Actually, there is a small probability that w_t acquires an arbitrary large value, i.e., in a finite period t a path can get off to a distance proportional to t, or even formally arbitrarily farther. At first sight, such hardly probable paths can be neglected. We show in Chapter 7, however, that paths of this kind can produce a principal contribution to advection of a scalar or vector admixture, thus giving rise to intermittency. Note also that the increment of the Wiener process principally differs from the increment of, e.g., a deterministic function $f(t) = t^{1/2}$ near $t = 0$. At any moment the Brownian particle has, so to speak, a new start and the increments are independent. Therefore, the derivative dw/dt is infinite not only at $t = 0$ but also at any other moment.

[c] Let us cite an exact form of the Wiener path for small periods, i.e., the recurrent logarithm law established by A. Khinchin in 1933: for almost all realizations (almost all ω) and for $t \to 0$, the upper and lower limits of the trajectory are given by

$$\overline{\lim}\, w_t = \left(2t \ln \ln \frac{1}{t}\right)^{1/2},$$

$$\underline{\lim}\, w_t = -\left(2t \ln \ln \frac{1}{t}\right)^{1/2}$$

respectively. Note that the function $\ln t^{-1}$ enters these expressions. It is understood that time is measured in units of the diffusion time L^2/κ, where L is the characteristic spatial scale, e.g., the size of a vessel filled with Brownian particles. In dimensional units, $\ln t^{-1}$ is replaced by $\ln L^2/\kappa t$. An exact value of L is unnecessary for the description of the trajectory at $t \to 0$ since $|\ln t| \gg \ln |L^2/\kappa|$ for $t \to 0$.

Owing to the Gaussian character of the Wiener process, it is completely determined by its correlation function. The latter can be determined through averaging of the following obvious identity:

$$w_t w_s = \frac{1}{2}[w_t^2 + w_s^2 - (w_t - w_s)^2].$$

Taking into account that $\langle w_t^2 \rangle = t$, $\langle w_s^2 \rangle = s$, and $\langle (w_t - w_s) \rangle^2 = |t - s|$ we obtain

$$\langle w_t w_s \rangle = \min(t, s). \tag{5}$$

What follows is an important property of the Wiener path: the increments $w_{t_1}, w_{t_2} - w_{t_1}, \ldots, w_{t_n} - w_{t_{n-1}}$ with $t_1 < t_2 < \ldots < t_n$ are independent since, according to (5), the average product of the increments vanishes. An increment of the coordinate during the Brownian motion over the period $t_2 - t_1$ is completely independent of all events which happened previously, at $t < t_1$. In addition, it follows from (5) that $-w_t$ is also a Wiener path. Obviously, the time stretching by a factor α is also admissible with simultaneous renormalization of the coordinate, i.e., $\alpha^{-1/2} w_{\alpha t}$ is also a Wiener path. The random process $t w_{t-1}$ is a Wiener path as well.

The Wiener random process is non-stationary: in the period t, a Brownian particle deviates by a distance of order $t^{1/2}$ in the sense of mean-square deviation. In the course of such motion, it occasionally returns to the origin in the one-dimensional case. In two dimensions, the returns are very rare while in a three-dimensional case, the Wiener process is not retraceable. These statements become obvious when we recall that fundamental solutions $G_{1,2,3}$ of the diffusion equation in one, two and three dimensions are given by

$$(2\pi t)^{-1/2} \exp\left(-\frac{x^2}{2Dt}\right), \quad (2\pi t)^{-1} \exp\left(-\frac{\rho^2}{2Dt}\right),$$

and

$$(2\pi t)^{-3/2} \exp\left(-\frac{r^2}{2Dt}\right)$$

respectively, where D is the diffusivity. The probability of return to a small vicinity of the origin for $t > t_1$ is given by $\int_{t_1}^{\infty} G_{1,2,3}(0, t)dt$. This integral diverges as a power-law on a line, as the logarithm on a plane and converges in three dimensions. Therefore, for $t_1 \to \infty$, returns become impossible in three dimensions.

To what extent does the Wiener path correspond to the motion of an individual Brownian particle? The non-differentiability means that the velocity is infinite at any point of a Wiener path. Hence, the acceleration is also infinite. Therefore, the Wiener paths describe the motion of particles with zero masses.

Indeed, let the mass of a Brownian particle be M and let it move under kicks from molecules each of mass m. At the moment of every impact the particle momentum abruptly changes, i.e., the velocity is described by a step function, while the force and acceleration are δ-functions of time. In the limit of vanishingly small molecular mass m and period τ between their impacts, the force is represented as a sum of a large number of uncorrelated δ-functions with small coefficients. Then the velocity has the Hölder derivative of the order 1/2 and the trajectory, correspondingly, has the derivative of the order of 3/2 (see Chapter 5). Next, when the particle mass tends to zero, $M \to 0$, the trajectory acquires the Hölder derivative 1/2, i.e., it becomes a Wiener process. The description of a particle with non-zero mass and finite τ can be accomplished using the Ornstein-Uhlenbeck process (see Section 4.6).

However, in some problems, e.g., in the analysis of diffusion of a scalar or vector admixture in turbulence, it is more convenient to consider the Wiener path as fictitious random paths. The real, physical meaning can then be ascribed only to the result of averaging the set of all such paths. The diffusing or spreading of a substance can be readily described in these terms. This approach has an analogy in quantum mechanics, where the wave function itself has no physical meaning and the physically observable quantities are derived by averaging the squared modulus of the wave function as a weight. Quite similarly, averaging over the Wiener path requires knowledge of their statistical weight, or in mathematical terms, the measure of Wiener process.

4.3. AVERAGING OVER PATHS

Consider a Wiener path which begins from $x_0 = 0$ at $t_0 = 0$ and comes to $x = x(t)$ at the moment t. Dividing the time axis into intervals $0 < t_1 < t_2 < \ldots < t_n = t$ we obtain the finite-dimensional vector $w_n = (w_{t1}, w_{t2}, \ldots w_{tn})$. Independence of the increments, $x_1, x_2 - x_1, \ldots$ and their Gaussian properties, imply that the distribution density of this vector is equal to the product of the Gaussian densities for the individual intervals:

$$P(t_1, t_2 - t_1, \ldots, t_n - t_{n-1};\quad x_1, x_2 - x_1, \ldots, x_n - x_{n-1})$$

$$= \frac{1}{\sqrt{2\pi t}} \exp\left\{-\frac{x_1^2}{2t_1}\right\} \frac{1}{\sqrt{2\pi(t_2 - t_1)}} \exp\left\{-\frac{(x_2 - x_1)^2}{2(t_2 - t_1)}\right\} \tag{6}$$

$$\equiv \prod_{j=1}^{n} \frac{1}{\sqrt{2\pi(t_j - t_{j-1})}} \exp\left\{-\frac{(x_j - x_{j-1})^2}{2(t_j - t_{j-1})}\right\}.$$

A continuous Wiener path is obtained in the limit of vanishing length of the intervals when the vector w_n, with $n \to \infty$, becomes infinite-dimensional. The corresponding limit of the distribution density (6) yields the Wiener measure of the trajectory w_t:

$$\mu^w = \int \prod_{j=1}^{\infty} \frac{1}{\sqrt{2\pi(t_j - t_{j-1})^1}} \exp\left\{-\frac{(x_j - x_{j-1})^2}{2(t_j - t_{j-1})}\right\}. \tag{7}$$

The measure (7) allows integration in infinite-dimensional functional space, i.e., evaluation of $\int F(w)\, d\mu^w$. In other words, the averaging of the functionals $F(w(t))$ can be accomplished. The idea of averaging of functionals with the Gaussian weights belongs to the French mathematician Levi (1972).

An outstanding contribution to the integration of functionals that arises in problems of radiophysics and statistical mechanics is due to Wiener (1958) who has formulated, in particular, the definition of the measure (7).

Those who have never dealt with integration over an infinite-dimensional functional space when evaluating the Wiener integrals could feel uneasy with all these seemingly complicated, demanding and number-decorated theorem procedures. Actually, in the Wiener integral the things are much simpler.

In the preceding chapter, we have noted two approaches if interpreting an integral. In the framework of Riemann's approach, an integral is the area under the line $y = f(x)$ and it can be calculated as follows. Divide the x-axis into many intervals and construct rectangles elongated along the y-axis whose combination approximates the considered area. The sum of areas of all rectangles can be obtained without difficulty and what remains is to take the limit for finer divisions of the x-axis. Another understanding of an integral belongs to Lebesgue who proposed to divide the y-axis rather than the abscissa. It would seem that the difference is inessential. However, now we do not need detailed knowledge of the properties of the x-axis (even when generalized to functional space), but it is sufficient just to be able to calculate the areas over this axis.

We have employed Lebesgue's idea when evaluating the integrals over such an abstract object as the space of elementary events. Let us now apply this idea to evaluate an integral in which the integrand depends on the functional variable. According to Lebesgue, all we should know is how to calculate the area, or what is the measure over which the integration in a functional space is carried out.

Let us recall how the area is understood on an ordinary plane. One should introduce the unit of length and define the area of a simplest object, e.g., a rectangle (the product of width and length). Next, the considered planar object, of however complicated shape, is approximated by a set of rectangles which allows us to calculate directly the approximate value of the area. What remains to be done is to decide what the measure of a rectangle's counterpart is in the functional space and the integrals over the functional space could be evaluated as easily as those over a straight line. Mathematicians are often blamed for excessive formalization, but appropriate formalization (Lebesgue's formalization is of this very kind) is constructive!

But how should one understand a rectangle in a functional space?

In order to formulate this concept[d] note that physically the considered integrals are averages of the functionals that depend on the trajectories of a Wiener process. In other words, the integrands are trajectories of a Wiener process. The "rectangles" should be some simple sets of these functionals, and their measures should be the probabilities of the trajectory of the Wiener process belonging to these sets.

To construct the desired sets, let us take the example of croquet, a game that is so popular in England. Consider the horizontal plane on which the various trajectories of the Wiener process are plotted with installed small croquet arches (hoops) parallel to the axis $x = W$. The set of hoops is determined by the moments t_i to which they correspond and by the coordinates of their left-hand and right-hand ends of the arches, a_i and b_i (see Fig. 4.1). The "rectangle" in the functional space, usually called the quasi-interval, is simply the set of functions whose plots pass through all the hoops, while the measure of the quasi-interval is defined as a probability that the trajectory of a Wiener process passes through the hoops, i.e., belonging to the quasi-interval. The specific positions of the hoops can be arbitrary, provided only that their number is finite. This is the very physical definition of the measure of a quasi-interval, or the Wiener measure, that is required for the practical evaluation of the Wiener integrals. Of course, the measure of the quasi-interval can be calculated explicitly, as was done in the derivation of expressions (6) and (7):

[d] The simplest procedure that comes to mind is to expand the functions into Fourier series and regard it as belonging to a cube in a functional space with those functions whose Fourier coefficients belong to the interval $(0, a)$. The value a then would be the side of the cube. However, it would be a non-rigorous extension of finite-dimensional concepts to an infinite-dimensional space. Indeed, the volume of such an N-dimensional cube is $V_N = a^N$. Let $N \to \infty$. For $a > 1$ we have $V_N \to \infty$, while $V_N \to 0$ for $a < 1$. It is clear that a reasonable definition of the measure cannot be formulated this way.

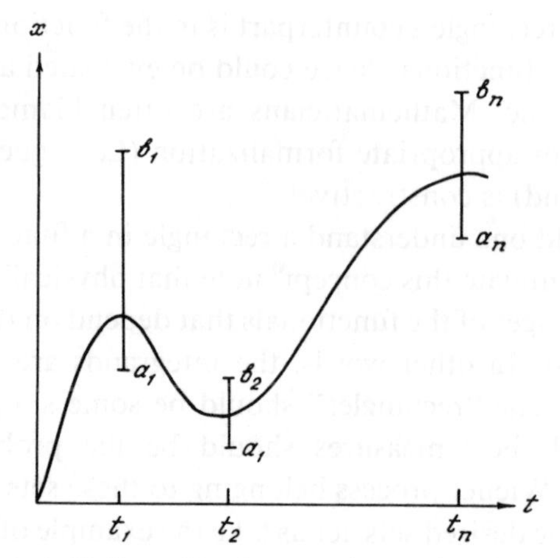

Fig. 4.1. The quasi-interval is formed by those continuous functions whose plots fit within wickets shown here.

$$\mu Q[t_1, t_2 - t_1, \ldots, t_n - t_{n-1};\quad x_1, x_2 - x_1, \ldots, x_n - x_{n-1}]$$

$$= \int_{a_1}^{b_1} dx_1 \int_{a_2-a_1}^{b_2-b_1} d(x_2 - x_1) \ldots \int_{a_n-a_{n-1}}^{b_n-b_{n-1}} d(x_n - x_{n-1})$$

$$P(t_1, t_2 - t_1, t_n - t_{n-1}; x_1, x_2 - x_1 \ldots x_n - x_{n-1})\,.$$

We notice that this is an integral variant of the expression (7).

4.4. THE WIENER AND FEYNMAN INTEGRALS

In the discussion above, we have intentionally hidden a subtlety which leads to many difficulties in quantum mechanics. Lebesgue's recipe for the construction of the measure[e] requires the possibility of using not only

[e] This is one of numerous examples of discoveries which are unrecognized and underestimated by their authors: for one-dimensional integration Lebesgue's measure represents purely mathematical importance; its principal physical importance was discovered in connection with functional integration.

finite but also infinite sets of rectangles approximating the considered geometric figure (more precisely, they are countable sets of rectangles). In turn, this requires the possibility of performing the following operation. It should be possible to divide any rectangle into a countable number of non-intersecting rectangles and the measure of the original rectangle should be equal to the sum of the measures of smaller rectangles (this property is known as denumerative additivity of the measure). It can be easily verified that the Wiener measure (7) obeys this requirement which is due to rapid decrease of exponential functions in (7) for the growing separation between the points x_j and x_{j-1}.

In quantum mechanics, Feynman's approach employs the averages over quantum wanderings rather than the averages over the Wiener paths. Formally, transition from a latter path to the quantum path consists in the replacement of t by it (compare the diffusion equation with the Schrödinger equation) but, as a result, the exponential functions in (7) cease to decay at infinity, and Lebesgue's approach becomes inapplicable. What is obtained then is another measure which has much weaker properties and is known as the Jordan measure. This measure does not ensure the existence (convergence) of the Feynman integrals used in quantum mechanics. In simpler terms, when using a Feynman integral, one can never be sure that it converges until it is calculated out as a certain finite number. In the case of Wiener integrals considered here, we are free from this difficulty: their convergence can be verified as easily as that of an ordinary one-dimensional integral. This is the very difficulty which is meant by mathematicians when they say that the functional integration is not sufficiently substantiated.

The typical functionals encountered in these applications are of the power-law or the exponential form. For example, in quantum mechanics, the transition amplitude (whose square is the probability) for transition from the point x_0 to x_n is given by the integral over the trajectories connecting these points:

$$K(x_0, x_n) = \int e^{iS(w)} d\mu^w ,$$

where the functional S is known as the action. This is essentially the integral over time from t_0 to t_n of the Lagrange function (Feynman and Hibbs, 1965).

A pure mathematician would be satisfied by pointing out that the Wiener integral is a correctly defined quantity. For a physicist, it is principally important to know whether he can calculate this integral to obtain the desired numerical estimates of observable quantities. Let us illustrate this procedure in the simplest case where the functional under an integral sign is a constant, say, unity. Any Wiener integral can be calculated in two different ways. The Wiener integral can be understood as an average or mathematical expectation M over the Wiener paths. Then, of course,

$$M1 = 1.$$

On the other hand, the Wiener integral can be understood as the limit of the iterated integral over the measure (7). Divide the time interval into N periods of the length Δt. Then

$$\int 1\, d\mu^w = \lim_{N\to\infty} \frac{1}{(\pi\Delta t)^{1/2}} \int_{-\infty}^{\infty} \ldots \int_{-\infty}^{\infty} \exp\left\{-\sum_{i=1}^{N-1} \frac{(x_{i+1} - x_i)^2}{\Delta t}\right\} dx_1 \ldots dx_N.$$

By transformation of variables $y_{i+1} = x_{i+1} - x_i$, the iterated integral reduces to the product of ordinary integrals of the type

$$I_w = \frac{1}{(\pi\Delta t)^{1/2}} \int_{-\infty}^{\infty} \exp\left\{-\frac{y^2}{\Delta t}\right\} dy. \tag{8}$$

All of them are equal to unity. We see again that the considered Wiener integral is equal to unity. This result has a more fundamental meaning than that which is evidenced in the simple example. The integral of unity is, of course, equal to the total volume (or area, or the measure) of the considered space. Thus, the measure of the total space of paths is equal to unity.

A similar calculation can also be done for Feynman's quantum integrals. In this case, suitable normalization of the measure by a constant A also gives unity in the result. However, not only the rapidly converging integrals of the type (8) are calculated at every step, but also the integrals

$$I_F = \frac{1}{A}\int \exp\left\{\frac{iy}{\Delta t}\right\} dy.$$

These integrals of rapidly oscillating functions also converge, but this convergence is only conditional and the result of calculation depends on the way of summation. This fact underlines again that the Feynman measure is much less productive than the Wiener measure (in formal terms, the former is not denumeratively additive and the integration region cannot be divided into a countable set of sub-regions during the calculation).

The difficulties associated with the definition of Feynman's integral are objective, rather than due to the laziness of mathematicians. Nevertheless, physicists have used these integrals widely and successfully. No less useful are the Riemann integrals that also are not based on denumeratively additive measure. Moreover, the physical mind is not dissatisfied with using not only conditionally convergent integrals but even integrals that have sense only as generalized functions. In all such cases, the difficulty is a consistent definition of the measure. This is avoided by considering only certain restricted class of integrands specific for every particular considered case. This is exemplified by Feynman's integrals that lead to quite reasonable results in known applications. Thus, there is some inadequacy of mathematical foundations which consists in the necessity to isolate and describe a reasonable class of integrable functionals, rather than in the formulation of the theory of measures.

Other Wiener integrals can be calculated similarly to the calculation of the integral of unity stated above. The reader can find these results in the review of Gelfand and Yaglom (1956). However, we do not consider it important to enlarge the number of examples where the Wiener integrals can be calculated explicitly. It is more important to find out whether there are problems which can be easily solved in terms of continual integrals but are hard nuts to crack for other approaches. When the language of continual integration was being developed, the answer to this question was not clear. On the one hand, Feynman's idea has allowed a deeper understanding of the least action principle. In particular, physicists disguised earlier the fact that trajectories of motion provide the minimum of action only at small time intervals, while at larger periods they correspond only to the stationary points of the action functional. (For a function of a single variable, $y(x)$, the necessary condition for the extremum is the vanishing of the derivative, $y' = 0$. For the functionals, the corresponding condition is the vanishing of the variation, i.e., the Euler equations. The sufficient condition for a minimum is a positive value of the second derivative at the stationary point, $y'' > 0$. The similar condition for the action functional is the so-called restricted Legendre condition, i.e., a positive value of the second derivative of the Lagrangian with respect to velocities; usually this condition is fulfilled automatically. However, in order to ensure a minimum of a functional, it is required to fulfill also the so-called Jacobi condition whose meaning reduces to the requirement for all possible trajectories, as described by the Euler equations for various initial velocities, not to intersect in the course of evolution of the system. Therefore, for a pendulum, the action is minimal only during periods not exceeding the oscillation period.) For the Feynman integrals, the action is stationary when the phase of the integrand is stationary. Those trajectories which correspond to the non-stationary action strongly interfere to attenuate the resultant contribution to the integral. Thus, the least action principle (or, more exactly, the stationary action principle) can be interpreted as the quasi-classical approximation for Feynman integrals. It turns out that there is no necessity to check a formidable condition of minimal action. We may see how sagacious was the Ukranian philosopher Grigory Skovoroda (1722–1794) who wrote that unnecessaries are difficult and necessaries are easy.

Now the apparatus of continual integration has become a working instrument in many fields. In Chapters 8 and 9, we will discuss some examples of physical problems where the solutions can be easily obtained with the help of Wiener integrals but are hardly accessible with other methods which would lead to cumbersome and unconvincing arguments.

One more aspect of calculations of Wiener integrals in the numerical approach is of special interest. Integrals extending along straight lines were already known to Archimedes, who had learned the integration of certain power-law functions. However, Archimedes could not have known such a powerful tool as the Newton-Leibniz formula that allows the reduction of definite integrals to indefinite ones. Far from it that all integrals can be evaluated analytically with this method; nevertheless, the desired result can be obtained in many very important cases. Without this method, Archimedes' calculation of even the simplest integrals becomes a complicated and tedious undertaking. The computer era has turned matters upside down; with computers, calculations of practically any integral along a straight line is an elementary problem. Had Archimedes been a contemporary of von Neumann and Turing, this so useful and usual analysis of Newton and Leibniz could never have appeared. Meanwhile, the numerical calculation of derivatives is by far more complicated than integration. The necessity to calculate ratios of small numbers leads to tremendous loss of accuracy which must be taken care of by, e.g., least-squares interpolation.

The Wiener integral can be also calculated numerically. The pioneering results in this direction are due to Cameron (1951). The most convenient approach considers the integral as an average over the Wiener measure rather than as a limit of an iterated integral. It is sufficient to consider three dozen typical Wiener paths and calculate the average over them – the central limit theorem guarantees that the result approximates the Wiener integral. Basically, this is the same idea as that of the Monte-Carlo method. As always, reality is not so benevolent. First, the construction of the Wiener paths requires generators of random numbers. Any software package contains something called a generator of random (more exactly, pseudo-random) numbers. Computational mathematicians have proposed many beautiful algorithms for such generators but the very formulation of the problem is intrinsically contradictory: a rigid sequence of operations, e.g., an algorithm, is expected to give something random. Of course, the situation is not desperate. The onset of chaos from order in strange attractors is now well studied, even if it is not completely understood, and widely used in programming the generators of random numbers. However, a light-hearted "first come, first served" approach to the choice of the algorithm can lead to impressive results. An algorithm which is useful in a certain class of problems may work absolutely unsatisfactorily in other cases. For instance, some algorithms that produce good samples of points randomly distributed along a line cannot be generalized to the three-dimensional case because they produce clusters around some planes or more complicated surfaces (see Fig. 4.2).

Another difficulty in the numerical calculations of Wiener integrals is associated with the fact that some of them are determined by unlikely events, by non-typical paths. Such situations are considered in Chapter 8. A similar difficulty is also known for ordinary integrals of rapidly oscillating functions. Special methods should be applied to them but the problem does not seem to be unsolvable.

These two kinds of difficulties are actually intrinsically connected: the more exotic the paths that make principal contributions to the Wiener integral, the tougher are the requirements that arise for the reproduction of these paths, i.e., eventually, for the used random numbers generator. Notwithstanding these difficulties, the simplest Wiener (and Feynman) integrals are within the reach of modern numerical methods.

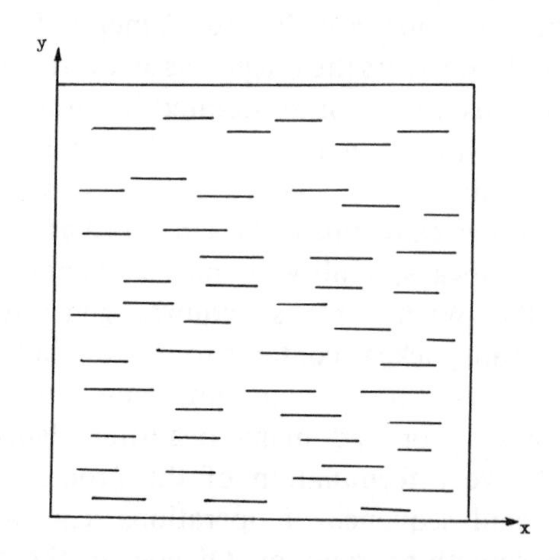

Fig. 4.2. A pseudo-random numbers generator, which satisfactorily produces random sequences x_n on a line, can yield characteristic structures on the plane when these numbers are combined into pairs $x = x_{2n}, y = x_{2n+1}$. This example refers to a standard generator of pseudo-random numbers FARN which is provided with many software packages. This illustration of its disadvantage has been published by Maximov (1986) in one of the Soviet popular journals which regularly discuss software for microcomputers. It was reported in one of the subsequent issues that omission of some of the generated numbers, e.g., the choice between $x = x_{5n}$ and $y = x_{5(n+1)}$, removes this correlation (Salzmann, 1987). Similar difficulties are mentioned by Jones (1988).

Discussing the justification of these methods, we inevitably touch upon a remarkable and surprising peculiarity of computational mathematics. This field of science is genetically connected with pure mathematics, the science which exists in the name of rigorous proof. In modern computational mathematics, however, even if only in some cases, irreproachable mathematical proofs can be given. They prove to be almost useless after a deeper analysis. As a result, modern computational mathematics is unexpectedly close to experimental physics in the nature of its arguments; in both cases, the principal instrument of verification is experimentation, either in a laboratory or numerical. Let us clarify this by the following example. When computing an ordinary integral, one replaces the integrand by a polygonal line or by using a more precise Simpson's parabolic rule. One should know how many mesh points must be taken in order to reach the desired accuracy. This problem was solved long ago theoretically and all calculus textbooks give the theorem with their corresponding estimates. The proof can be made as rigorous as possible for theorems of calculus. However, the result itself is practically useless. At least, nobody thinks of incorporating it into standard software. The reason for this is simple: the derived estimate of the required fineness of the mess depends on the maximum derivative of the integrand. For other interpolations, even higher derivatives are involved. The computation of these is much more complicated than the computation of an integral. Therefore, instead of this rigorously proved estimate, the so-called Runge's empirical estimate is usually applied, which is based on the presumption that the result of the computations is close to the true value of the integral when the doubling of the number of meshpoints does not lead to a considerable change in the result. This estimate cannot be proved while the examples that directly contradict it can be given explicitly. Nevertheless, in the hands of a judicious person, this empirical estimate gives excellent results.

4.5. THE WHITE NOISE

The force $\mathbf{F}(t)$ that appears in Section 4.1 is a random, rapidly varying function with zero average value, $\langle \mathbf{F}(t) \rangle = 0$. Thus, the particle momentum is also zero on average. Suppose that the force is constant over time intervals of length Δt and its values in these periods are independent. Let

Δt tend to zero. In order to transmit the finite r.m.s. momentum to a Brownian particle, the force should be proportional to $(\Delta t)^{-1/2}$. Then for any two moments t and t', the mean of the product $F(t)\,F(t')$ vanishes if $|t - t'| > \Delta t$ and is infinitely large if $t = t'$ (as dictated by the finiteness of the transmitted r.m.s. momentum). A compact expression of this has the form

$$\langle \mathbf{F}(t)\,\mathbf{F}(t')\rangle = \delta(t - t')\,, \tag{9}$$

where the force amplitude is taken to be unity for convenience. The correlation function (9) has a flat Fourier spectrum since

$$\int_{-\infty}^{\infty} \langle \mathbf{F}(t)\,\mathbf{F}(t+s)\rangle \exp\{-i\omega s\}\,ds = 1\,.$$

A random process with a spectrum of the correlation function which is flat and independent of the frequency ω is usually called a white noise. All frequencies are equally represented in such a process.

Of course, the white noise can never be realized in nature because a real random process always has a non-vanishing characteristic correlation time (memory) τ. However, the correlation function (9) serves as a very convenient mathematical idealization of a process whose memory is short (as compared with all other characteristic times). More exactly, this is the limit of a short-correlated process for $\tau \to 0$. This limiting process proves to be directly connected with the Wiener process.

To elucidate this connection, consider the integral of the random force, i.e., the momentum $\boldsymbol{\pi}(t) = \int_0^t \mathbf{F}(s)\,ds$. Due to non-correlation of the force at the moment t and that at $t' > t$, the quantities $\boldsymbol{\pi}(t)$ and $\boldsymbol{\pi}(t') - \boldsymbol{\pi}(t)$ are statistically independent. In addition,

$$\left\langle \int_0^t \mathbf{F}(s)\,ds \right\rangle = 0\,,$$

$$\left\langle \int_0^t \mathbf{F}(s)\,ds \int_0^{t'} \mathbf{F}(u)\,du \right\rangle = \int_0^t \int_0^{t'} \langle \mathbf{F}(s)\,\mathbf{F}(u)\rangle\,du\,ds = t' - t\,.$$

Thus, the momentum has all the properties of the Wiener path, and hence

$$\mathbf{F}(t) = \frac{d\mathbf{w}_t}{dt}. \tag{10}$$

Thus, the white noise is the derivative of the Wiener process. But $\mathbf{w}_t$ does not have an ordinary derivative. Therefore, the relation (10) has only a symbolic, generalized meaning. In other words, explicit operations with the process $\mathbf{F}(t)$ and the equations which include it are meaningless. Meaningful are the average value and the correlation function of the white noise. The corresponding equations must be understood only in the integral sense. Nevertheless, Langevin and his followers, including the most high-brow mathematicians, use explicitly the white noise and differential equation with it, agreeing with the symbolic meaning of the corresponding procedures and appreciating their brevity and convenience.

The probability distribution of the white noise is Gaussian, as well as the measure of the Wiener process. This fact can be perceived intuitively starting from the following property of Gaussian distributions. If the sum of two independent random quantities is Gaussian, both of them are also Gaussian (see, e.g., Hida, 1980). The reverse statement is not only true but can even be very easily verified with the use of the explicit expression for the Gaussian distribution.

4.6. THE ORNSTEIN–UHLENBECK PROCESS

Let us note a more smooth random process as compared with the Wiener process. This process has a finite correlation time and a steady distribution, in contrast to the Wiener process whose trajectories with probability one eventually go away to infinity.

The Ornstein–Uhlenbeck process $\mathbf{v}_t$ is given by the solution of the stochastic Langevin equation with the white noise as random force,

$$\frac{d\mathbf{v}_t}{dt} = -\frac{\mathbf{v}_t}{\tau} + \sigma^{1/2}\frac{d\mathbf{w}_t}{dt}. \tag{11}$$

By integration one obtains

$$\mathbf{v}_t = \mathbf{v}_0 e^{-t/\tau} + \sigma^{1/2} \int_0^t \exp\left\{-\frac{t-s}{\tau}\right\} d\mathbf{w}_s . \tag{12}$$

For constant (or normally distributed) v_0 the process is Gaussian with the expectation value and dispersion as follows:

$$\langle \mathbf{v}_t \rangle = \mathbf{v}_0 \exp\left\{-\frac{t}{\tau}\right\},$$

$$\sigma = (\mathbf{v}_t - \langle \mathbf{v}_t \rangle)^2 = \frac{\sigma\tau}{2}\left(1 - \exp\left\{-\frac{2t}{\tau}\right\}\right).$$

For $t \to \infty$, these characteristics tend to zero and $\sigma\tau/2$, respectively. The latter values determine the abovementioned steady Gaussian distribution. The correlation function can also be deduced directly from (12). For large t and s and for finite difference $t - s$, it is given by

$$\langle \mathbf{v}_t \mathbf{v}_s \rangle = \frac{\sigma\tau}{2} \exp\left\{-\frac{|t-s|}{\tau}\right\}. \tag{13}$$

Note that in this limiting case the correlation function depends only on the time lag $|t - s|$. Such processes are called *stationary*. We see that the values of v are considerably correlated only at the moments t and s that are separated no larger than τ; the latter has the meaning of the correlation time. For $t \to 0$, the function (13) vanishes for all t and s except for $t = s$ when it reduces to the delta function provided $\sigma = \tau^{-2}$. In this sense, the Wiener process can be considered as the limiting case of the Ornstein–Uhlenbeck process under decreasing correlation time.

CHAPTER

FIVE

FRACTALS AND DIMENSIONS

In the preceding pages, the reader has already encountered the mathematical concepts (e.g., the Lebesgue integral) beyond the scope of university courses for physicists. We cannot say that these concepts are more complicated than the traditional ones: an integration after Lebesgue is not more difficult than that after Riemann. These concepts were not only required for solving problems which physicists faced a few decades ago but they also provide clearness in solving many modern problems. In the following chapters, we continue to employ modern mathematical physics. The crucial moment which changed the physical and mathematical consciousness and marked the beginning of wide acceptance of these novel ideas was when Mandelbrot (1977) published his book in which many facts of traditional mathematics were exposed at a novel, clear and constructive level. These ideas were not properly comprehended by mathematicians, nor were they known to physicists, but lay collecting dust in libraries. The ideas wandered from one textbook to another in the role of mathematical subtleties, inaccessible to uninitiated scholars. Nowadays, these ideas should become a working tool for physicists. Ignorance of the motion of a fractal will soon be considered as equal to a lack of skill in differentiation or integration.

This chapter is devoted to fractals and fractional dimensions.

5.1. HISTORICAL PERSPECTIVE

"Geometricians say that the line is a length without width; and we sceptics, cannot conceive a length lacking width, neither in perception nor in speculation" (Sextus Empiricus, Against the Physicists, 1,

390–391). Even in ancient times (the quotation refers to 2nd century BC), people felt that the concept of dimensionality as inevitably an integer is restricted. Step by step, the concept of an object which is more voluminous than a line but still akin to a line, was being formed (a line having width). Very significantly, for a long period of time, this work of thought was wrapped in negative forms, i.e., in the forms of criticism of the concepts of one- and two-dimensional objects. This attitude develops vividly in the above quotation from Sextus Empiricus.

The concept of a line has played a decisive role in the formulation of analytical geometry. One of the branches of this field is represented by differential geometry. This line of thought goes beyond the analysis of lines and surfaces in our native and dear three-dimensional space. The development of differential geometry has led to the concepts of curved, non-Eulerian space which can be of many dimensions and of a curved complex of space and time.

Probably, yet a more important role in the development of science was played by calculus — differential and integral. However queer the forms acquired by some formal and abstract definitions of derivatives and integrals, the discovery of these ideas is intrinsically associated with the concepts of motions and curves (the derivative identified as a velocity or a slope and the integral as distance or area). It is important to note that what was meant here were smooth motions and curves. It is hardly probable that Newton and Leibnitz could arrive at the idea of a derivative if Brownian motion had been studied, i.e., velocities and trajectories of microscopic particles in thermal motion.

Newton, Leibnitz and many others right up to L. Euler took for granted the smoothness and existence of derivatives. This suppression was not passed on to the following generations. Textbooks have been filled by phrases like "derivative, provided it exists . . .". Like road signs warning a driver, these reservations have a deep meaning. It can be argued at which stage of education that such reservations and complications should be introduced. The successful development of the so-called non-standard analysis (Cartier, 1984; Shubin and Zvonkin, 1984), which uses freely the idea of infinitely large and infinitely small members, has revealed how this criticism sometimes carried too far and shows that the theory of limits can be exposed more closely to the approaches of Newton and Leibnitz than is usually done. Nevertheless, the very fact of the existence of unsmooth functions and related objects is not being called into question.

In the second half of the 19th century, K. Weierstrauss and G. Peano proposed examples of functions which are continuous and nonetheless have no derivatives and of curves which densely fill a square everywhere. From a modern viewpoint, properties of these objects seem strange because they are considered one-dimensional while it is more natural to endow such objects with higher dimensionalities, including fractional ones. Modern terminology calls them fractals. This preliminary stage of study of fractals was unconstructive. Their connections with physics and other possible applications were not appreciated and the very term "fractal" appeared relatively recently. The contemporaries often interpreted the activity of the critics as destruction of mathematics; thus, Ch. Hermite wrote to T. Stieltjes in 1893 that he "turned away in fear and horror from this lamentable plague of functions with no derivatives" (Hermite, 1905). The following generation of critical-minded mathematicians have analyzed equally intently the very foundations of their science — the concepts of the set, the natural number, the proof, etc. The reaction to this attack was expressed by D. Hilbert (1925) who hoped that nobody could expel us from the paradise created by Cantor.

The positive and constructive exploration of non-classical objects and situations was started by a theoretical physicist, P.A.M. Dirac, who was not a mathematician and was working in another field. His idea of delta function has made a tremendous impression for both mathematicians and physicists. His work has resulted in the reconsideration of the results of critical direction aimed at the formulation of useful, constructive ideas. Mentioning the influence of Dirac's work on the development of the theory of fractals, we mean primarily his conceptual influence, the change of scientific atmosphere. The point is that the concept of dimension belongs to the kind of everlasting problems of science which are reconsidered again and again in every new scientific era and for which changes of attitudes and approaches are no less important than the solution of a new particular important problem.

From the viewpoint of modern science, a function which lacks a derivative is far from being simply an abstract notion insidiously mentioned at any calculus examination: it is a trajectory of the Brownian particle. Due to its extremely ragged shape, such a trajectory should be considered as a "thick" line, or a fractal. As we shall see below, the very description of fractals is very much similar to the example of nowhere-differentiable function given by Weierstrass. It turns out that Weierstrass unconsciously had at hand the concept of the fractal! The arsenal of mathematics has provided an analytical apparatus suitable for describing such unsmooth objects. The notion of dimensionality is extended to fractional dimensions introduced by Hausdorff (1918) while the derivative is generalized by the so-called Hölder index, or the fractional derivative (this concept is due to the work of many mathematicians).

The idea of fractional dimensions is based on the analysis of the concept of integer dimensions. We easily understand what it means when we hear that a line or a circumstance is one-dimensional, a plane or spherical surface is two-dimensional, and a space or solid sphere is three-dimensional, etc. Roughly speaking, this means that the position of a point on a line is described by a single coordinate, that on a plane by two, and that in space is characterized by three numbers. A quantity such as the number of coordinates, of course, cannot be fractional. The introduction of fractional dimensions requires two steps: firstly, to make certain that the traditional understanding of dimensions is incomplete, in such a way that endorsement of fractional dimensions to some object would be a remedy; secondly, a certain definition of dimension must be given that allows for fractional values (Zeldovich and Sokoloff, 1985).

This plan can be accomplished as follows. One-dimensional objects are associated with length, two-dimensional ones with area, and three-dimensional with volume. The corresponding combinations of units of measuring dimensions (similar roots are not accidental in these words) are cm, cm^2 and cm^3, respectively. Experience with other fields of physics tells us that dimensions can involve fractional exponents. For instance, the dimension of an electric charge[a] in cgs is $g^{1/2}\ cm^{3/2}\ s^{-1}$. In order to implement such fractional dimensions into the fractal theory, the notion of coordinate should be considered in a somewhat broader sense. A point

[a] Of course, these fractional exponents should not be considered as an indication of the fractal nature of electric charge: in the natural system of units ($\hbar = c = G = 1$) the dimension of the electric charge has integer exponents (Synge, 1960, Appendix B) ($[e]$ = sec.). Nevertheless, there exist certain physical objects whose dimensions have fractional exponents in the natural system of units; all known objects of this kind are fractals (Barenblatt and Zeldovich, 1971).

inside a square can be characterized to any given precision by one coordinate, rather than by two, provided the corresponding coordinate line fills the square sufficiently densely. Actually, coordinates of this kind are far from exotic. For instance, the address of a townsman could, in principle, have the form of geographical longitude and latitude of his apartment, or even his bed. However, we choose another way of indicating the street, the house and apartment number; these are similar to the integer and fractional parts of the coordinates. In principle, all apartments could be labelled with consecutive numbers (the address system in Tokyo and the ZIP code system closely approach this idea). Now the question about the dimensionality of a town does not seem to be casuistical: when considered as combinations of streets, towns are one-dimensional; as regions on the surface of the earth, they are two-dimensional; recalling that houses can have elevators we would say that the relevant number is three. But can the true dimension be fractional? In other words, how should a town be characterized — by the total length of the streets (a sum of the first powers of lengths of small street segments separated by crossroads), or by its area (a sum of squared lengths of the abovementioned segments in towns of rectangular lay-out), or by any other powers of these lengths, e.g., three halves, dictated by a particular question of interest or a particular kind of town? The answer to this question is discussed in detail at the end of Section 5.2.

A difficulty with the idea of fractional dimension, as well as with the integer dimension, is due to the non-uniqueness of the way of its introduction discussed in the last two paragraphs. Other approaches can be proposed, which are equally natural but lead to other results. In the next section we begin our discussion with another approach which is less familiar to physicists but was proposed earlier than the others. This approach, proposed by Hausdorff, is based, in fact, on ideas formulated already by Euclid. The essence of this approach is the recognition that one- and two-dimensional objects are essentially parts of a three-dimensional space whose one or two dimensions are very small.

To conclude this historical introduction, we should mention two basic ideas associated with the concept of fractals. When discussing a ragged and unsmooth trajectory of the Brownian particle and the corresponding infinitely large acceleration, we understand that this is an idealization. On a very small scale finiteness of the particle mass and of the time between collisions become essential and the trajectory's smoothness is recovered. When discussing a physical fractal surface, we visualize it as a rough surface where the height of irregularity slowly decreases with decreasing area of its projection. However, these irregularities should be much bigger than interatomic distances; otherwise the very notion of the body surface would be inapplicable. When describing how a long molecule fills a three-dimensional volume (the corresponding mathematical object is known as the Peano curve), we understand that beginning with a certain degree of filling up, we should take into account not only the linear dimension of the molecule but also its thickness and our description of a tangled curve becomes inapplicable. Therefore, the concept of fractals is generically associated with the concept of intermediate asymptotics (Barenblatt and Zeldovich, 1976; Barenblatt, 1978): the size of irregularities is small but still much larger than a certain smallest scale.

The second idea can be classified using a rough surface as an example. Imagine a surface with two or more families of irregularities whose base areas are the same but for the height probability distributions. On the other hand it is also conceivable that the surfaces are irregular only in the narrow vicinity of some point. It is clear that both these cases cannot be described by a single number or even by a fractional dimension. The notion of fractals is based on the presumption that forbids such situations: the corresponding spectral expansions should have random phases of individual terms. Analysis of these situations requires either an independent investigation, as in the theory of delta function, or an introduction of the hierarchy of dimensionalities corresponding to every subsystem of irregularities (see Section 5.6 below).

We would like to mention that the era that gave birth to the theory of fractals (19th century) was followed by the era of criticism in mathematics (20th century) when many other strange objects were discovered. In contrast to fractals, these still have no applications. These are such monsters as the extraordinary sets whose elements are they themselves, non-Aristotlean logics, intuitional mathematical analysis, etc. Possibly, they will eventually find their places in physics. The reader who wishes to ponder over this theme is referred to Fraenkel and Bar-Hillel (1958), one of the rare monographs in this field whose language is accessible to non-specialists.

5.2. FRACTALS: THICK LINES

Let us examine the working edge of a razor blade under a microscope, tuning the latter to finer and finer resolutions. To the naked eye the edge seems to be smooth; increasing the resolution, we see irregularities and notches. Further, they seem larger, and finally a crystallic structure is seen. We enter the quantum world are the edge can no longer be precisely defined. The concept of the fractal is devised for the description of the intermediate range when the edge loses its smoothness but the interatomic level of the resolution is still far away (Fig. 5.1).

We have already mentioned that one of the first concrete mathematical examples of fractals was given by Weierstrass (1886): a function without a derivative. This function has one-to-one correspondence with the x-axis and is given by the series

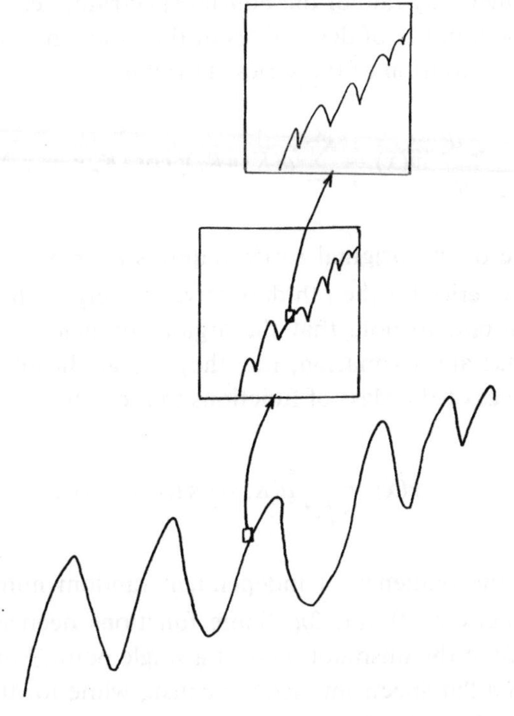

Fig. 5.1. A fractal curve has a self-similar structure over a wide range of scales.

$$y(x) = \sum_{n=0}^{\infty} A^n \cos(B^n \pi x),$$

where $O < A < 1$ and the product AB is sufficiently large (Weierstrass chose $AB > 1 + 3/2$ while a more natural condition $AB > 1$ was established by Hardy (1916)). It is convenient to consider from the beginning a more general form,

$$y(x) = \sum_{n=0}^{\infty} a(K_n) \cos(K_n x + \varphi_n), \tag{1}$$

where $a(K_n) \propto K_n^{-\alpha}$, $K_n \sim n \to \infty$.

Weierstrass' example can be easily recovered from (1) for the vanishing phases φ_n and partial summation. It is well known (see, e.g., Zygmund, 1965) that the decay rate of the Fourier spectrum, i.e., the value of α, is related to the number of derivatives of the function $y(x)$. Indeed, term-by-term differentiation of the series (1) yields

$$\frac{d^j}{dx^j} y(x) = \sum_{n=0}^{\infty} a(K_n)(K_n)^j \cos(K_n x + \varphi_n).$$

Convergence of the original series requires $a_n \propto K^{-\alpha}$ with $\alpha > 0$. It is clear that the series for the j-th derivative converges when $\alpha > j$.

It is important to note that the arguments below are based on the random phase approximation, i.e., they are applicable only to typical representatives of the class of functions that can be written as

$$y(x) = \sum_{n=0}^{\infty} a(K_n) \cos(K_n x + \varphi_n),$$

where φ_n is the sequence of independent random numbers uniformly distributed between 0 and 2π. Some functions decrease their spectra gradually due to the unsmoothness at a single point. For instance, delta function has a flat spectrum, $a(K) = \text{const.}$, while for the step function $a(K) = \alpha^{1/K}$; these functions have irregularities only at the origin.

Meanwhile, fractal curves have a structure such that, statistically, all their points have the same properties. Note also that K_n and n are not equal but only of the same order of magnitude. In other words, it is important that, together with the phases, the frequencies of the harmonics are also uncorrelated. Otherwise, for small values of α (specifically, $\alpha < 1/2$) the plot $y(x)$ can acquire randomly distributed infinitely high peaks. To illustrate these subtleties, consider as example the function $y = \Sigma_{n=1}^{\infty} \sin(nx)/n^{\alpha}$ with $\alpha < 1/2$ and $0 \leqslant x < 2\pi$. Firstly, phases of different harmonics, nx, are equal (and thus, completely correlated). Secondly, $K_n = n$ for this function. Due to these facts, behaviour of this function is dominated by two peaks at $x = 0$ and at $x = 2\pi$: firstly, the integral $\int_0^{2\pi} y^2(x)\,dx$ diverges; secondly, the series can be shown to converge uniformly (and, thus, the function is continuous) everywhere except at the points where $\sin(x/2) = 0$; however, we see that $y = 0$ at these very points, e.g., $x = 0$ and $x = 2\pi$. The only possibility is the presence of singularities in the vicinity of the points $x = 0$ and $x = 2\pi$.

In modern terms, the principal part of Weierstrass' paper consists in verifying the fact that the function constructed by him is typical, or in other words, generic. However, the very notion of generic properties was formulated more recently, after Cantor proposed his theory of sets. This notion has made a significant impact on the value of mathematical labour: in the 19th century people spent years proving that, say, the number π is irrational while now we are usually quite satisfied by observation that practically all numbers are irrational.

Thus, it follows from the theory of trigonometric series that for $0 < \alpha < 1$ the function $y(x)$ given by (1) is continuous but non-differentiable. The plot of this function is a fractal line. For $1 < \alpha < 2$ the "path" $y = y(x)$ is smooth but the corresponding velocity is not smooth at every point, i.e., it has fractal properties. Only those functions have all non-fractal derivatives (like the functions $(1 + x^2)^{-1}$ or $\exp(-x^2)$) whose spectrum $a(K)$ decays faster than any degree of K (e.g., as $\exp(-K)$ or $\exp(-K^2)$).

All the results formulated above are negative because the ideas of calculus have only restricted applicability to functions with slowly decaying spectra. The introduction of the Hölder index allows to cast the results in a positive form and say that the function $y(x)$ given by (1) has smaller-than-one derivative, specifically the α derivatives. For the

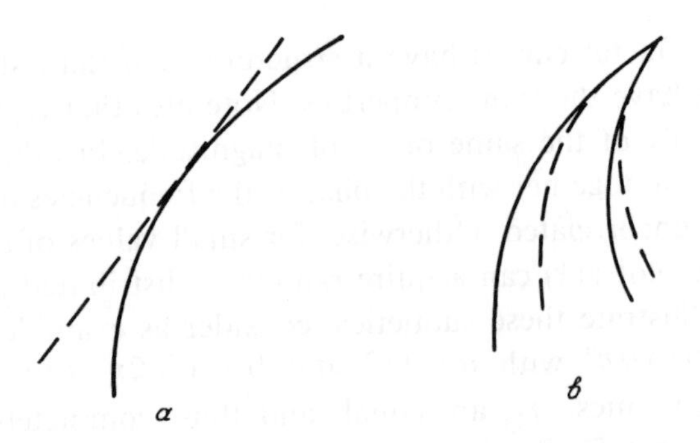

Fig. 5.2. The structure of smooth and fractal curves in the vicinity of some point (a): a smooth curve has a tangent line (dashed); (b): a fractal curve has a curvilinear cone $\Delta y = |\Delta x|^{\alpha}$ as the tangent (dashed).

common derivative we have, by definition

$$\Delta y = \mu \Delta x.$$

For our function we have (see Fig. 5.2)

$$\Delta y = \mu_H (\Delta x)^{\alpha}, \tag{2}$$

where μ and μ_H are the ordinary and Hölder derivatives, respectively. More exactly, for $\Delta x < 0$ and $\Delta x > 0$ the left-hand and right-hand derivatives, μ_{H_-} and μ_{H_+} respectively, should be introduced. After this, manipulations with the function $y(x)$ of (1) become nearly as easy as with smooth functions.

Two examples of physical objects with deficient smoothness are widely known. One of them is the Wiener process, i.e., mathematical idealization of the Brownian motion (see Chapter 4) with the particle mass and the time between collisions tending to zero. In this limiting case, the particle experiences an acceleration of the delta-function type at any moment and the correlator of the Wiener process has a flat spectrum. A sum of many instantaneous kicks produces the displacement $w_{t+dt} - w_t = w_{dt} = dw_t \sim (\Delta t)^{1/2}$, i.e., the corresponding Hölder index is $1/2$: the Brownian particle trajectory has half a derivative. Itô (1946) has proposed a remarkable formula for the evaluation of the differential of a smooth function of the Weiner stochastic process, $F(w_t)$. The differential dF, i.e., the increment with accuracy up to dt, is given by $dF = F'dw_t + (1/2)F''(dw_t)^2$ rather than $F'dw_t$. It is the second term in Itô's formula which formally allows the development of an alternative description of diffusion (after Langevin-Smoluchowsky-Wiener), not based on the diffusion equation but later arriving at expressions of the Kac-Feynman type for solutions of evolutionary equations through the Wiener integrals (or the Feynman integrals in quantum mechanics). This method and its applications to the description of heat conduction and diffusion will be discussed in more detail in Chapter 8.

Another example is provided by the Kolmogorovian turbulence. If we neglect viscosity that becomes essential at small scales, turbulent pulsations of the velocity field are of the order $\delta v \sim (\delta r)^{1/3}$. This implies that although the Kolmogorovian velocity field is continuous, it has only 1/3 of the spatial derivative. Here the physical object has a more complicated nature: this is an unsmooth vector field.

Let us now calculate the dimensionalities of fractal curves which were discussed above. After Hausdorff, the fractional dimension of a set of points can be determined as follows. Surround every point of the set by a circle of small radius ε and calculate the area of the combination of all circles, $S(\varepsilon)$ (in fact, for any given ε a finite number of circles would suffice). When some circles overlap, the area shared by two or more circles is counted only once. The rate of decrease of $S(\varepsilon)$ with ε determines the dimension. Indeed, for a smooth curve $S(\varepsilon) \sim \varepsilon L$, L being the curve length. When the curve degenerates into a point, we have $S(\varepsilon) \sim \varepsilon^2$. In the other limiting case, when the considered set occupies part of a plane, $S(\varepsilon) \sim \varepsilon^0$, since the number of circles declines as ε^{-2}.

For the fractal curve (1), the covering area is estimated in the following way (Fig. 5.3). For the wave vector $K = \varepsilon^{-1}$, individual periods of the sinusoid find room within the circles of the radius ε. If the coefficients $a(K_n)$ decrease with ε (remember that $K_n \sim \varepsilon^{-1}$) slower than ε for $\varepsilon \to 0$, the width of the circle band is of the order of $a(\varepsilon^{-1})$ rather than ε. If, in contrast, $a(\varepsilon^{-1}) = 0(\varepsilon)$, then the circle band can manage to follow all the bands of the curve, which is smooth in this case. Thus, for $\alpha < 1$, the dominant contribution to the area of combined circles come from those harmonics which have $K_n \sim \varepsilon^{-1}$. Let the length of the projection of the

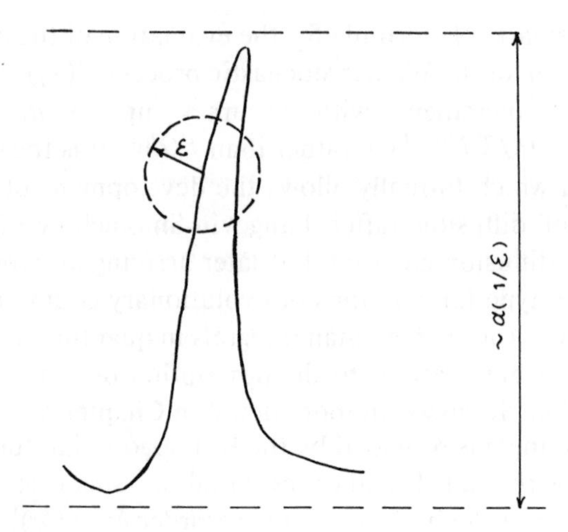

Fig. 5.3. Definition of the external dimension of a fractal curve: In the fractal case the width of the curve vicinity is determined not by the radius of the circle but by the magnitiude of bendings of the curve.

considered segment of the x-axis be of the order of unity. The combination of all circles has area $S(\varepsilon) \sim \alpha(\varepsilon^{-1}) \sim \varepsilon^{\alpha}$. For $0 < \alpha < 1$ the area decreases with ε slower than for a smooth curve and the corresponding fractal curve is intermediate between a line and a surface. Hausdorff has proposed a definition according to which the dimension of such an object is given by

$$\dim_{\text{ext}} \gamma = 2 - \alpha. \tag{3}$$

The subscript "ext", which is an abbreviated "external", indicates that when measuring this quantity we have penetrated outside the curve itself and γ symbolizes the considered curve.

A serious shortcoming of the discussion above is the exceptional role of the x-axis. It can be easily modified to make all axes equivalent. This is reached through a parametric description of the curve γ, i.e.,

$$x = \sum_{n=0}^{\infty} a(K_n) \cos (K_n t + \varphi_{1n})$$

$$y = \sum_{n=0}^{\infty} b(K_n) \cos (K_n t + \varphi_{2n}).$$

Both directions are equivalent when the coefficients a and b decay similarly:

$$a \sim b \sim K_n^{-\alpha}, \quad 0 < \alpha < 1.$$

Estimation of the area of ε-vicinity of γ in two dimensions again yields ε^{α} (rather than $\varepsilon^{2\alpha}$ since now one should consider the deviations of irregularities from a certain average position). We again obtain $\dim_{\text{ext}} \gamma = 2 - \alpha$.

There is another approach to the definition of the fractal dimension of a line which is based only on its properties intrinsically associated with the curve itself rather than on the orientation of the curve γ on the plane. This approach was already mentioned in Section 5.1. Let us introduce the parameter t that varies along the curve γ. Divide the t-axis into intervals of length ε and calculate the length of γ taking into account only those bands in which t has an increment no less than ε. The total length of such segments is of the order of $a(\varepsilon^{-1})(1/\varepsilon) \sim \varepsilon^{\alpha-1}$ and tends to infinity with decreasing ε. We stress that this kind of analysis relies only on the possibility of measurement of a distance along the curve for a certain range of t, rather than the degree of tangleness of the curve.

It is worthwhile to ponder over the fact obtained above of the divergence of the length with decreasing ε. Assume that we have made an error in the preliminary estimation of the dimensionality of the considered object while actually we have at hand not a line but a part of a plane and now we attempt to parametrize a two-dimensional region with a single parameter. Of course, such parametrization is very unfortunate: the coordinate curves more and more, densely filling the region with self-crossings producing a kind of lattice (Fig. 5.4). The transverse period of the lattice is $\sim \varepsilon$ while the number of cells is $\sim \varepsilon^{-2}$. The total sum of lengths of the broken line segments is $\sim \varepsilon \cdot \varepsilon^{-2} \rightarrow \infty$. This sum diverges and here this only means that the region is two-dimensional and one should consider its area rather than its length, i.e., up-squared segment lengths rather than the lengths themselves.

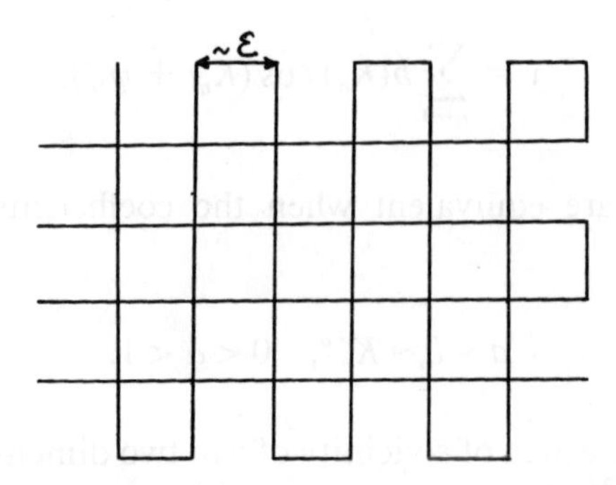

Fig. 5.4. Definition of the internal dimension of a fractal curve: Segments of the broken line, which densely cover the plane, are small but very numerous. The sum of the squared lengths of the segments is finite.

In the case of fractional dimension, the line is so tortuous that the sum of its elementary lengths is infinite and to obtain a finite result, one should sum up their fractional powers rather than the first or second ones. The resulting sum is finite (of course, if the initial guess of the dimension is correct) when this fractional exponent is $1/\alpha$. The dimension of this sum is $\mathrm{cm}^{1/2}$ and the number $1/2$ itself represents dimensionality. Thus, the internal dimension of a line can be naturally defined as

$$\dim_{\mathrm{int}} \gamma = \frac{1}{\alpha}. \tag{4}$$

Of course, this definition is valid for lines embedded in space of any dimensionality.

The external dimension of planar fractal curves vary from 1 to 2 (the dimension of a plane), while the internal dimensions range from unity to infinity. These two dimensions coincide only in the trivial case of smooth lines. Evidently, every specific physical problem requires the use of a properly chosen kind of fractal dimension. For instance, in the problem of absorption by a thin tangled thread, one estimates the number of atoms that can find place in its vicinity, i.e., the external dimension is relevant. However, where the thread weight is concerned, the internal dimension is of importance.

Let us calculate the external dimension of a line specified by (1) which is embedded in three-dimensional space. In this case

$$x_i = \sum_{\substack{n=0 \\ i\leqslant 3}}^{\infty} a_i(K_n) \cos (K_n t + \varphi_{in}), \quad a_i(K) \sim K^{-\alpha}.$$

Estimation of the volume of ε-vicinity of the line clearly yields $V(\varepsilon) \sim \varepsilon^{2\alpha}$. By analogy with the two-dimensional case, we have

$$\dim_{\text{ext}} \gamma = 3 - 2\alpha. \tag{5}$$

for three dimensions. For m dimensions

$$\dim_{\text{ext}} \gamma = m - (m - 1)\alpha.$$

Hence, the external dimension of a fractal line lies between the dimension of the smooth line and the dimension of its surrounding space, while the internal dimension is between the smooth line dimension and infinity.

5.3. FRACTALS: THICK SURFACES

The definition of the fractal dimension of a rough surface can be introduced in a similar way. Let the surface ϕ be specified parametrically by

$$x_i(u, v) = \sum_{\substack{K_u \\ i=1,2,3}} \sum_{K_v} a_i(K) e^{i(K_u u + K_v v)},$$

$$a_i \sim |K|^{-\alpha}.$$

Since now we have a double Fourier series, its sum is finite when α exceeds unity. Therefore, in the considered fractal range $0 < \alpha < 1$, the double series diverges in an ordinary sense and it can be summed up only in the principal value sense. In other words, the harmonics should be summed up within a finite sphere first and then summation in radii should be accomplished.

Let us divide the parameter space (u, v) into triangles of diameter ε. Division into squares is easier to interpret but division of into triangles (triangulation) is technically more feasible since, in contrast to a square, a triangle is a rigid figure, i.e., it is uniquely determined by the lengths of its sides.

Now we proceed to the calculation of the total sum of the areas of the triangles that approximate the surface. Again we take into account only those harmonics whose wavelengths are not shorter than ε. This sum is of the order of $[a(\varepsilon^{-1})(1/\varepsilon)]^2(1/\varepsilon^2)$ and for $0 < \alpha < 1$ it diverges with decreasing ε. The sum becomes finite when $1/\alpha$ powers of areas are summed up and the surface dimension is given by

$$\dim_{\text{int}} \Phi = \frac{2}{\alpha}, \quad 0 < \alpha < 1 . \tag{6}$$

As for the external dimension, it is given, after Hausdorff, by

$$\dim_{\text{ext}} \Phi = 3 - 2, \quad 0 < \alpha < 1 , \tag{7}$$

since the volume of ε-vicinity of the surface is of the order of $a(\varepsilon^{-1})(1/\varepsilon)$.

The fact that external and internal dimensions of fractal objects do not coincide implies that, in contrast to regular multiple differentiable objects, for fractals the link between internal and external characteristics is broken. In other words, in the case of a regular surface, the curvature can be calculated by two methods: either with the help of the Riemann curvature tensor, as is done in general relativity theory or with the help of curvatures of cross-sections, and both results coincide. Meanwhile, for fractal surfaces, the results would generally differ. It is remarkable that such "fractal" behaviour is also typical of smooth but irregular surfaces, i.e., surfaces that have tangent planes but not higher derivatives, even though the dimension of such surfaces is exactly two. This interesting phenomenon, discovered by the American mathematician J. Nash, leads to curious situations. Consider first a regular surface of area S. For a closed surface, it is natural to expect that the volume enclosed by the surface is of the order of $S^{3/2}$. Of course, the numerical factor depends on the particular shape of the surface and on the distribution of the Gaussian curvature, but the order of magnitude is correct. Nash (1954) has discovered that when the surface smoothness or the number of Hölder derivatives is between 1 and 2 (i.e., when its external curvature still cannot be evaluated through the usual formulae), its surface area is not connected with the enclosed volume: a crumpled sphere (with preserved metric, i.e., when distances between all points can be measured along the surface) of arbitrarily large surface area can be placed within an arbitrarily small sphere. According to modern estimates, the exact threshold smoothness at which this phenomenon appears lies between 1.07 and 1.7 (Bakelman *et al.*, 1974). Nash's surfaces demonstrate the effect anticipated by the Czech writer Jaroslav Hasek who puts into his character's mouth the conjecture that, inside the earth, there is another sphere, much larger than the outer one. The simplest conception of dimension is widely used in biology for the explanation of the proportions of various creatures like corals or sponges. It would be interesting to find out the species that demonstrate the properties associated with Nash's character of their bodies.

5.4. FRACTALS: A FOAMY SPACE-TIME

The concept of internal dimension can be extended without significant modifications to curved spaces, e.g., the general relativity (of course, in this case the space should be divided not into triangles but into their four-dimensional counterparts – simplexes). If on microscopic scales the space-time resembles a foam, as suggested by Wheeler (1962), then the dimensionality to a macroscopic observer can differ considerably from that observed microscopically. This and other related problems are still poorly investigated by mathematicians but some estimates can already be done.

Cosmology frequently invokes the so-called flat spectrum of perturbations in which all harmonics of the density perturbation in the Friedmann cosmological model have equal probability. Clearly, such a flat spectrum can be considered only as an intermediate asymptotic regime.

Initially, this spectrum was proposed as a convenient reasonable assumption in the absence of more definite information (Zeldovich and Novikov, 1967). However, it became clear recently that the so-called inflation theory leads to a cosmological model with Euclidean comoving space and flat spectrum of density perturbations (Starobinsky, 1982; Halliwell and Hawking, 1985). In this case the curvature also has a flat spectrum. This means that in the scale range of the intermediate asymptotic, the comoving space dimension is not exactly three but slightly larger.

At a later stage of evolution the fractal properties of the space are lost. The cellular structure that arises from this (Shandarin and Zeldovich, 1984) is structured matter distribution rather than structured space. Of course, Einstein's equation predicts that singularities of the metric arise at positions where matter density diverges due to the presence of the cellular structure. These singularities are similar to those arising when two non-coaxial cylinders intersect (Zeldovich *et al.*, 1977). However, the spectral expansion of such a metric is strongly phased rather than fractal.

The multi-dimensional Nash surfaces have new interesting properties. When the dimensionality of the surrounding space is sufficiently large, surfaces can possess Nash's properties even if it is differentiable arbitrarily many times (Nash, 1956). For the four-dimensional surfaces, the threshold dimension of the enveloping space, corresponding to the appearance of such crumpled surfaces, lies somewhere between 11 and 29 (Gromov, 1970). In this connection, it is proper to recall multi-dimensional generalizations of the field theory of the Kaluza-Klein type, which are believed to be far from the theory of fractals. In these models the micro-scale dimension of space is larger than the macroscale dimensionality because some dimensions turn out to be periodic coordinates whose period is infinitesimal. Another interpretation of this reduction of dimensionality at macro scales was also proposed (Joseph, 1962). In multi-dimensional space-time, potential can be introduced which is similar to the potential that confines quarks within a nucleon but prevents the existence of free particles outside a four-dimensional space-time surface rather than the existence of free quarks. This picture implies a certain connection between the external and internal properties of surfaces: the volume of the vicinity of a surface should be of the order of $V^{N/n}$ where V is the n-dimensional volume of the surface, n is its dimension and N is the enveloping space dimension. It seems that the presence of Nash's crumpled surfaces in the enveloping space would make this picture impossible, i.e., the dimension of the enveloping space exceeds the value between 11 and 29. The lower bound of this interval is in intriguing closeness to the dimension of the enveloping space predicted by the particle theory (Cremmer *et al.*, 1978).

5.5. FRACTALS: CONTOUR LINES

The initial approach to the concept of fractals in physics was associated with a striking example – measurement of the Great Britain coastline length. Recall that the coastline can be introduced as the solution to the equation $h(x, y) = 0$, where $h(x, y)$ is the height above sea level.

It turned out that the larger the scale of map used, the longer the coastline. The coastline turns out to be a fractal whose measurement requires the introduction of a "quasi-length" of the dimension of centimeter raised to a certain power. Although specialists are well aware of these circumstances, the practice of cartography still neglects it. As an example, we mention that the administration of the National Park of Lithuania publishes the area, depth and coastline length of the Park's lakes. The last can hardly be reasonably defined (even though easily determined by ignorants); meanwhile, the lakes' diameters are not cited although this quantity can be defined consistently. Of course, the length of the coastline can be used in practice but necessarily with reference to the corresponding scale, which requirement is rarely observed.

The definition of fractals which was introduced above does not exactly correspond to this classical problem. The point is that detailed inspection of the coastlines reveals the system of skerries which add many relatively shorter segments to the total coastline length.

In contrast, the considered parametric fractal objects can have self-intersections (e.g., two-dimensional trajectories of the Brownian type). Meanwhile, the contour lines have no self-intersections with probability unity because the conditions for simultaneous vanishing of a random function and its gradient, required for contour line self-intersection, lead to an over-determined system of equations.

In order to determine the fractal dimension of the contour line of a random function, one should find out how many Hölder derivatives it has.

This number, α, depends on the behaviour of the spatial correlator at small distances. For a smooth random field,

$$\langle \varphi(\mathbf{x})\varphi(\mathbf{x}+\mathbf{r})\rangle = 1 - Ar^2$$

for $r \to 0$, while for a fractal field

$$\langle \varphi(\mathbf{x})\varphi(\mathbf{x}+\mathbf{r})\rangle = 1 - Ar^\alpha .$$

The fractal dimensionality of the contour line is given by the index α.

5.6. FRACTALS: DENSE POINT SET

All examples of fractals considered above have no small dimensionality. The objects of dimensionality, say, 10^{-1} arise when one considers a set of a large (infinitely large in the limiting case) number of points that densely fill the fractal object. In order to determine the Hausdorff dimensionality of such planar fractal, every point should be surrounded by a circle of the radius ε, the area of the combination of all circles should be calculated, and its dependence on ε should be determined. Such fractals have been known for a long time in mathematics and exemplified, e.g., by the Cantor set[b] (an example of a set that is unconnected everywhere and has the power of continuum). However, such objects did not seem to be more than the fruits of mathematicians' mind games. Now we know that sets similar to the Cantor set are attracting sets (strange attractors) of dynamic systems (see Section 5.8); the set of zeros of one-dimensional Wiener process is also fractal (of dimensionality 1/2). Recently, fractal objects of this type were proposed for the description of intermittency in turbulence (Frisch, 1983; Parisi, 1984). We have already mentioned that in the Kolmogorov turbulence the velocity is continuous but has only 1/3 derivative. This fact can be expressed graphically as follows. Let us plot along the abscissa the number σ of derivatives of the velocity field, and

[b] By construction, the Cantor set is the following. Divide the unit-length interval into three parts and discard the middle one. Repeat this procedure with every sub-interval. The limiting set that arises after an infinite number of steps is called the Cantor set. It consists of points rather than of intervals.

along the ordinate the Hausdorff dimensionality $d \equiv d_{\text{ext}}$ of that set of points at which the velocity has the given number of derivatives (Fig. 5.5). Then for the Kolmogorov turbulence we have

$$d(\sigma) = 3\theta\left(-\frac{1}{3} + \sigma\right),$$

where $\theta(x)$ is unity for $x > 0$ and zero for $x < 0$. Thus, for $\sigma < 1/3$ the prescribed differentiability is observed almost everywhere, while for $\sigma > 1/3$ such differentiability can be found practically nowhere. The function $d(\sigma)$ of this form corresponds to the Gaussian velocity field, i.e., to absence of intermittency.

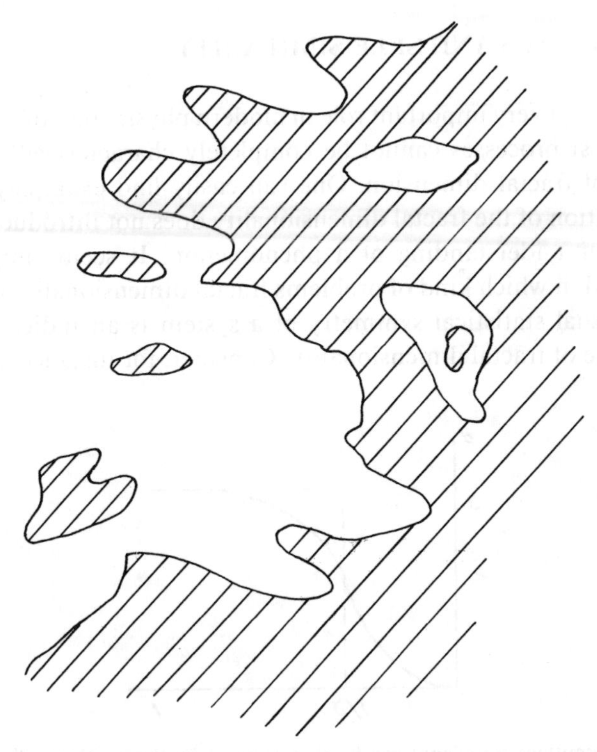

Fig. 5.5. The schematic structure of level lines of a random function. Dashed lines mark the parts below "sea level". The island structure, "skerries", are visible.

When there are coherent structures present in a turbulence and it is essentially non-Gaussian, it can be expected that the value of the function $d(\sigma)$ continuously changes from three to zero (Frisch, 1983). Practically, it is more convenient to measure in an experiment a certain relation for variations of higher velocity moments with its distance and its order (Parisi, 1984). Such measurements made by Anselmet *et al.* (1984) have allowed us to determine the function $d(\sigma)$ for a real turbulence shown in Fig. 5.6. Accuracy of this determination is still rather low but some deviations from Gaussian properties are clearly seen. Qualitatively, this plot implies that the velocity field has 1/10 derivative practically everywhere and 9/10 derivative almost nowhere; for intermediate orders there occurs a smooth change of dimensionality of the set of points where the velocity field has a given smoothness.

5.7. FRACTALS AND SELF-SIMILARITY

Fractals play a very important role in modern physics but, of course, even the simplest processes cannot be completely characterized solely by a quantity of fractal dimension. One can easily find examples when the determination of the fractal dimensionality does not introduce anything new to our understanding of a phenomenon. It seems important to understand in which kind of problems fractal dimensionality is informative. Unusual statistical symmetry of a system is an indication of the importance of fractal dimensionality. Consider, for instance, a dense set

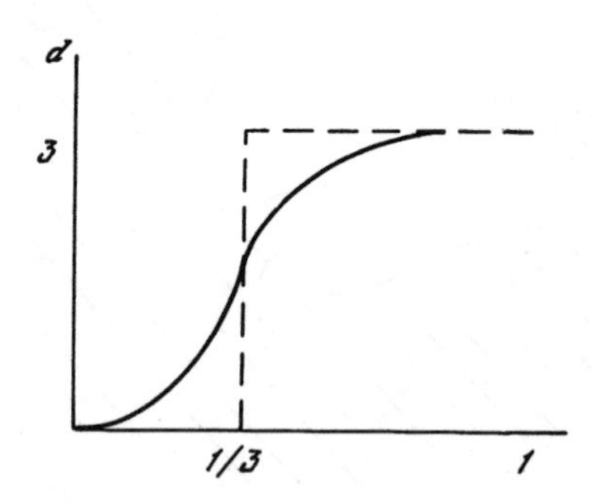

Fig. 5.6. Intermittent turbulence can be characterized by the fractional dimension. The dashed line shows the dimension characteristic of the Kolmogorov turbulence; the solid line refers to the real turbulence.

of points with fractal properties, the so-called fractal cluster. In this case statistical symmetry can be understood as follows. Let us construct around every point of the cluster identical surfaces (e.g., equal spheres) whose size is considerably larger than the size of the points (the particles) but considerably smaller than the cluster size (the condition of intermediate asymptotics). Then the property of self-similarity implies that the number of points (or, the cluster mass or fraction, which is the same) are the same on average within every sphere and proportional to a certain power of the sphere radius. Such fractal clusters are exemplified, probably, by a globular lightning (Smirnov, 1986). At any rate, these fractal clusters are single objects that would provide a light weight and rigid structure for the active matter in the globular lightning. Within a globular lightning the air is heated and convective flows take place. The whole cluster structure can be considered as a system of pipes and the lift force produced by convection can be evaluated similarly to that produced by smoke moving above a chimney (Zeldovich, 1937). The detailed structure of the fractal cluster that forms a globular lightning is still unclear. If the cluster is produced by sucessive addition of individual particles in Brownian motion, its external Hausdorff dimensionality is approximately 2.46 (Smirnov, 1986). In this case this external dimensionality is of primary interest since it characterizes the relation between an empty space and the volume filled by active matter. A similar cluster in two dimensions has the dimension of about 1.71. In particular, self-similarity implies that the overall shape of the cluster is nearly isotropic.

Similar ideas work also in elementary particle physics, particularly in the description of the multiple particle production. The process of multiple particle production of hadrons is considered as the production of a large number of partons, i.e., quarks and gluons, with their subsequent combination into hadrons. In some approximate descriptions, development of this process is represented as the Brownian motion of partons between the positions where the particles are formed. Quite naturally, the opening angle b of the diffraction cone that results is proportional to the number of steps of the process of multiple particle production, i.e., to the average multiplicity of particles, $\bar{n}$:

$$b \sim \bar{n}.$$

Such dependence is not observed in experiments, which can be explained by the interaction between partons or by their rescattering. This makes the diffusion of partons more complicated and their diagrams resemble fractal trees π with many branchings. Here the most relevant characteristic of a fractal is its internal dimension which reflects a complexity of parton evolution, the tangled pattern of the diagrams. It can be shown (I.M. Sokolov, 1986) that in this case

$$b \sim \bar{n}^{-2/\dim_{\rm int}\pi},$$

where $\dim_{\rm int}\pi$ is the internal dimension of the fractal π. An experiment confirms this power-law dependence to first approximation and gives (Dremin, 1987)

$$\dim_{\rm int}\pi \approx 7.5,$$

which indicates that parton diagrams are highly tangled (note that $\dim_{\rm int}$ π considerably exceeds the dimensionality of surrounded three-dimensional space).

The role of self-similarity in the theory of fractals is so important that many early works in this field (see, e.g., Barenblatt, 1978) base the very definition of fractal dimensionality on the concept of self-similarity. Now such an approach seems to be an exaggeration; the Hausdorff dimension exists for many non-self-similar objects as well, even if its role is not so prominent in such cases.

As we repeatedly noted in this chapter, the fractal, or fractional dimensionality, is not a single concept but rather a group of related concepts. In various fields of science, various aspects of fractal dimensionality appear relevant. We illustrate this in detail using as example the application of fractal dimensionality to strange attractors which are used in modern physics as simple models of turbulence. This application will be useful for our discussion in Chapter 6.

5.8. STRANGE ATTRACTORS

This section and the following one are devoted to a particularly interesting example of fractal which bears an unusual name, the strange attractor (Ruelle and Takens, 1970). This fractal is useful for the description of chaotic behaviour of trajectories of some deterministic dynamic systems.

It is well known that the state of a dynamic system can be completely specified by positions and velocities, or $(\mathbf{x}, \dot{\mathbf{x}})$ in symbolic vectorial representation. The set of these quantities forms the phase space. The temporal evolution of the system is described by a certain trajectory in the phase space. The concepts of the phase space and phase trajectory were proposed and effectively used by Gibbs in statistical mechanics, by Poincaré in the theory of differential equations and by Birkhoff in the theory of dynamic systems.

According to a classical theorem (see, e.g., Nemytsky and Stepanov, 1952), through every point of the phase space $(\mathbf{x}, \dot{\mathbf{x}})$ of a system, for which

$$\frac{dx_i}{dt} = f_i(x_j), \quad i, j = 1, 2, \ldots, N$$

with continuous derivatives df_i/dx_j on the right-hand side, can pass only one trajectory of the system. Two or more trajectories can only approach asymptotically (be attracted to) for $t \to \infty$ a certain point, line, or another set which is called the attractor.

For instance, the phase plane of an undamped linear oscillator[c]

$$\dot{x}_1 = x_2; \quad \dot{x}_2 = -\lambda x_2 - \omega^2 x_1$$

is covered by elliptic trajectories that correspond to particular values of the energy $E = x_2^2/2 + \omega^2 x_1^2/2$. In order to place the system on a given ellipse, the corresponding initial conditions $(x_1(0), x_2(0))$ should be prescribed. In the presence of dissipation, $\lambda > 0$, the energy is not conserved and every trajectory becomes a spiral converging to a single equilibrium point (0, 0), which is an attractor of zero dimensionality. The damping of oscillations can be prevented by proper supply of energy. In a grandfather's clock for example, such supply is accomplished

[c] Physicists are more used to another form of description of the oscillator: $\ddot{x} + \lambda\dot{x} + \omega^2 x = 0$. We put $x = x_1$, and $\dot{x} = x_2$.

through the potential energy of a weight and the ratchet. The phase portrait of such a nonlinear oscillatory system features a special line or limit cycle which all trajectories approach. This is a classical attractor of dimensionality one discovered by A. Poincaré and was introduced into a theory of oscillations by A. Andronov (1929), an outstanding Soviet scientist who worked in Gorky for a long time and has created there his scientific school.

The revival of widespread interest in phase portraits of dynamical systems occurred in the sixties because of the problem of convection considered in a meteorological context. Studying the problem of convection, B. Saltzman (1962) has proposed and analyzed a finite-dimensional system of equations that described convection. He discovered that in some cases all components of the solution, apart from a certain three, decay in time while those three experience irregular undamped oscillations. Having properly appreciated this fact, E. Lorenz (1963) proposed the system of only three equations (two velocity components and the temperature deviation). Varying the principal parameter of the system (the Rayleigh number), he revealed that, apart from the static solutions and regular oscillations (which correspond to convective rolls of regular shape), the system possesses also non-periodic solutions of irregular and stochastic character. This result was quite unexpected because the Lorenz system looks rather simple:

$$\dot{x} = \sigma(y - x), \quad \dot{y} = -y + Rx - xz, \quad \dot{z} = -bz + xy. \tag{9}$$

Lorenz has revealed the irregular undamped solutions for $b = 8/3$, $\sigma = 10$ and $R = 28$. It was shown later that "stochasticity" arises in a certain order rather than a wide range of parameters (see the review of Gaponov and Rabinovich, 1979). For instance, system (9) was considered as a model of global activity of the Sun for $\sigma \to 1$ and large R (of the order of $(\sigma - 1)^{-1}$) (Ruzmaikin, 1981).

For the values of b and σ chosen by Lorenz, the behaviour of the system in the phase space (x, y, z) is as follows (see details in Afraimovich *et al.*, 1977; D. Yorke and E. Yorke, 1979). For $R < 1$ only the static solution $(0, 0, 0)$ is stable. For $R > 1$, three stationary solutions exist, $(0, 0, 0)$ and $(\pm[b(R - 1)]^{1/2}, \ \pm[b(R - 1)]^{1/2}, \ R - 1)$, i.e., the origin and two symmetrically situated points at the height $z = R - 1$. For $R < R_1 =$

13.926 all trajectories converge to one of these fixed points while for $R > R_1$ there exists an infinite number of periodic trajectories and irregular, turbulent trajectories that do not converge to any fixed point or periodic trajectory. For $R > R_2 = 24.74$, all three fixed points are unstable and for $R = 28$, a stable chaotic regime is observed (for $R \leqslant 50$) in which the trajectories are attracted to a certain manifold in the phase space.

The global properties of a dynamic system can be obtained when there is a bundle of trajectories. Such a bundle can be prescribed by a smooth initial distribution of points in the phase space and one may follow the evolution of its shape in the course of approach to the attractor (Fig. 5.7).

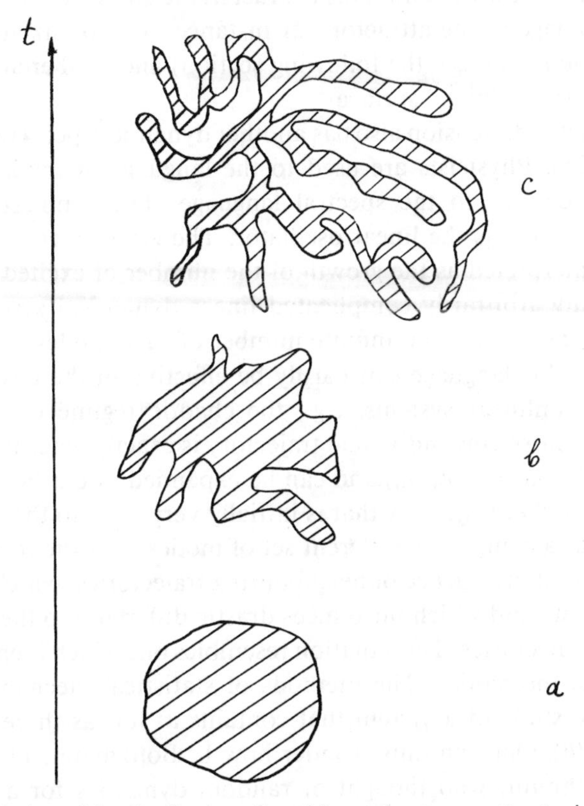

Fig. 5.7. Evolution of the distribution of particles at the strange attractor in phase space: initially the particles are distributed in clumps (a) but with time their distribution becomes more and more resolved (b, c).

Since divergence of the vector field determined by the system (9) is negative,

$$\frac{d\dot{x}}{dx} + \frac{d\dot{y}}{dy} + \frac{d\dot{z}}{dz} = -(\sigma + b + 1),$$

the attractor must have a zero measure (zero volume). On the other hand, the attractor asymptotically includes all trajectories that correspond to complicated three-dimensional motions. The basic rough characteristic of such unusual manifold is the dimensionality, which takes fractional values for strange attractors. This characteristic allows us to envisage the geometric image of the attractor; for instance, the Lorenz attractor has the dimension 2.09 (see the following section) and is, therefore, similar only to a curved "thick" surface.

The attractor dimension also has another dynamic aspect (Eckman and Ruelle, 1985). Physicists are used to the language of modes, e.g., the Fourier modes, or to the spectral language. The simplest harmonic oscillatory system is the linear oscillator. The growing complexity of a system is interpreted as the growth of the number of excited oscillators (modes). Any arbitrarily complicated linear dynamical system can be represented as a sum of an infinite number of such modes.

However, this language can hardly be effective in the description of essentially nonlinear systems, e.g., the chaotic regime of the Lorenz system. Of course, any individual trajectory is exactly determined by the initial conditions (x_0, y_0, z_0) and can be expanded into a Fourier series. However, another trajectory that is initially very close to the former one normally has a completely different set of modes. It is the instability, or the exponential divergence of neighbouring trajectories, which gives rise to stochasticity and which introduces drastic difference to the spectra of individual trajectories. The situation resembles one which is encountered in statistical mechanics. The methods of statistical mechanics can be used for the study of a system that contains as few as three equations (Lücke, 1976). One can only wonder how L. Boltzmann, J.C. Maxwell and J. Loschmidt, who thought of random dynamics for a system of $3 \times 6\ 10^{23}$ equations, consider this situation?!

The dimension of the strange attractor can be interpreted as the effective number of dynamic variables (dynamic modes) in the system. In this sense, it is said that the systems with strange attractors, i.e., with intrinsically stochastic dynamics, are low-modals. This idea can be of paramount importance for the problem of hydrodynamic turbulence. If it could be shown that a turbulent system has an attractor of finite dimensionality, this would give the number of basic modes that determines the flow properties and a simple description of turbulence would be possible. At any rate, according to Ruelle and Takens (1970), the transition from the laminar state to the turbulent one occurs through the excitation of a small number of modes. The earlier scenario of Landau and Hopf considers the excitation of an infinite number of modes (Landau and Lifshitz, 1959).

5.9. MEASURING THE ATTRACTOR DIMENSION

As in the case of any other fractal object, the dimension of an attractor can be determined, in principle, geometrically by covering it with the elementary cubes or spheres whose size is ε and whose dimension coincides with that of the phase space. The minimal number of cubes grows as ε^{-d} for $\varepsilon \to 0$ and the quantity d is called the external, or Hausdorff, or topological dimension of the attractor (see Section 5.2).

However, this idealized and mathematically rigorous procedure has two practical shortcomings. The first one is evident: it is technically difficult to determine d for a system with sufficiently large (> 3) dimension. The other shortcoming is more subtle: the described procedure does not distinguish any particular points of the attractor even though as a rule a trajectory approaches certain places more frequently than other places. Therefore, the exposure of the whole attractor set takes an impractically long time.

Another measure which is sensible to inhomogeneity of the passing frequency of a trajectory can be the entropy defined as follows.

Let us identify the trajectory $\mathbf{x}(t)$ with the microstate of the system measured with accuracy ε. Divide the phase space into cells of characteristic size ε. Let P_i be the probability that the trajectory $\mathbf{x}(t)$ passes through the i-th cell. Then, by analogy with statistical physics, the entropy of the system is given by

$$S(\varepsilon) = - \sum_{i=1}^{N(\varepsilon)} P_i \ln P_i, \tag{10}$$

where $N(\varepsilon)$ is the total number of cells. When decreasing the value of ε the entropy changes logarithmically:

$$S(\varepsilon) \simeq S_0 - d_s \ln \varepsilon, \quad \varepsilon \to 0, \tag{11}$$

the constant d_s is called the entropy or the information dimension (Famer, 1982).

It turns out that the entropy dimension can never exceed the topological one (Grassberger and Procaccia, 1983). Indeed, consider the set of successive values of the coordinate, $x_1, x_2, \ldots, x_k, \ldots, x_m$. The probability of finding x_k within the i-th cell is

$$P_i = \lim_{m \to \infty} \frac{\mu_i}{m},$$

where μ_i is the number of positions x_k that are within this cell.

For homogeneous covering of the attractor, $P_i = N^{-1}$. As follows from (10) and the equality $N = N_0 \varepsilon^{-d}$,

$$S(\varepsilon) = \ln N(\varepsilon) = \text{const.} - d \ln \varepsilon.$$

Thus, for homogeneous covering we have $d_s = d$. It is well known that under any deviation from homogeneity, the entropy only decreases. Therefore, $d_s \leqslant d$.

Grassberger and Procaccia (1983) have introduced the attractor dimension which is very convenient for practical purposes. This dimension characterizes the shape of the spatial correlation function of the attracting set and can be found from successive measurements of one of the dynamic variables based on a computer simulation or experiment.

The very possibility of describing a complicated attracting set in multidimensional phase space using only one variable (i.e., a single projection!) is surprising. Such a possibility is associated with, first, the fact that dynamic variables that determine the coordinates of the phase space obey the system of ordinary differential equations, $\dot{x} = f(x, y, z, \ldots)$, $\dot{y} = g(x, y, z, \ldots)$, $\dot{z} = h(x, y, z, \ldots)$, ... which can be reduced by differentiation in time to a single equation that includes only $x(t)$ and its time-derivatives, $\dot{x}(t)$, $\ddot{x}(t)$, $\dddot{x}(t)$, ... Secondly, Ruelle (1981) has shown that instead of the sequence $(x, \dot{x}, \ddot{x}, \ldots)$ one can consider the sequence

$$x(t), x(t+\tau), x(t+2\tau), \ldots, x\big(t+(N-1)\tau\big) \tag{12}$$

characterized by a fixed lag τ. At any moment t_i the point sequence (12) belonging to the attractor can be conveniently written as

$$\{x_j\} = x_1, x_2, \ldots, x_i, \ldots, x_N.$$

Next we construct the vectors

$$\xi_i = (x_i, x_{i+1}, \ldots, x_{i+m-1})$$

whose lengths are m and which are constructed at every point $\mathbf{x}_1, \mathbf{x}_2, \ldots, \mathbf{x}_N$. Their length m should exceed the expected number of dynamic variables.

The spatial correlations are characterized by the following function:

$$C_i(r) = \frac{1}{N} \times \left\{ \begin{array}{l} \text{the number of points whose distance} \\ \text{to the } i\text{-th point does not exceed } r\text{:} \\ |\xi_j - \xi_i| \leq r \end{array} \right\}.$$

After summation over all the points, we come to a correlation integral of the form

$$C(r) = \frac{1}{N^2} \left\{ \begin{array}{l} \text{the number of point pairs whose} \\ \text{separation } |\xi_j - \xi_i| \text{ does not} \\ \text{exceed } r \end{array} \right\}.$$

More formally,

$$C(r) = \lim_{N \to \infty} \frac{1}{N^2} \sum_{\substack{i,j=1 \\ i \neq j}}^{N} \theta(r - |\xi_i - \xi_j|), \tag{13}$$

where θ is the unit step function and the point separation can be understood in the usual Euclidean sense:

$$|\xi_i - \xi_j| = [(x_i - x_j)^2 + (x_{i+1} - x_{j+1})^2 + \cdots + (x_{i+m-1} - x_{j+m-1})^2]^{1/2}.$$

Grassberger and Procaccia (1983) have determined the dependence of log C on log r for a number of well known dynamic systems which possess strange attractors and revealed that for sufficiently small r this dependence is a power-law:

$$C(r) \sim r^{\nu}, \quad r \leqslant \varepsilon << 1.$$

The power index determines the correlation dimension through

$$\gamma\nu = \lim_{\substack{N\to\infty \\ r\sim\varepsilon}} [\log C(r)/\log r]. \tag{14}$$

It is clear by definition that

$$\nu \leqslant d_s \leqslant d.$$

This important conclusion is justified by mathematical proofs and illustrated by explicit examples (Eckman and Ruelle, 1985). For example, for the mapping

$$x_{n+1} = ax_n(1 - x_n), \quad a = 3.5699456\ldots, \tag{15}$$

which possesses a strange attractor, calculations with the use of expressions (13) and (14) for $N = 30.000$ yields $\nu = 0.500$ (± 0.005) while $d_s \simeq 0.517$ and $d \simeq 0.538$. For the Lorenz system with $R = 28$, $\sigma = 10$, $b = 8/3$ and $N = 15.000$, the result is $\nu = 2.05 \pm 0.1$ and $\nu = \sigma = d$ with good accuracy.

It should be noted that log C does not always vary linearly with log r. For a very small r one usually encounters a considerable scatter of points associated with poor statistics (insufficiently large N) or experimental noise. For a large r, the finite size of the attractor becomes essential.

Therefore, for the practical application of this method of determining the dimension one should single out a linear segment of the plot (log C, log r) and measure the slope angle.

The first application of the method of Grassberger and Procaccia considered experiments with the Rayleigh-Benard convection; the result was $\nu = 2.8$ (Malraison *et al.*, 1983). The gradual increase of the dimension of the vector $\xi_i(m)$ shows that the measured dimension ν becomes stable when $m = 6 - 8$. Thus, the low numbers of modes excited in connection are directly indicated by the experiments. For an infinite effective number of modes, ν would grow infinitely with m. This fact was verified by the analysis of the white noise where ν proved to be growing linearly with m.

The system usually encountered in these experiments are nonlinear systems of a low number of excited modes contaminated by statistical noise, e.g., the white noise. Since noise dominates at small scales, for very small values of r one should obtain $\nu \sim m$ while for moderate r, the dimension ν tends to a certain, not very large, value as m grows; and $\nu < m$ always (Fig. 5.8).

To conclude this chapter, we note that the analysis of mathematical objects always has two different aspects. The first is the creation of the mathematical image of reality. For instance, the Peano curve is the mathematical image of a tangled thread ball. On the other hand — and this is the other aspect — Peano has constructed a specific example of

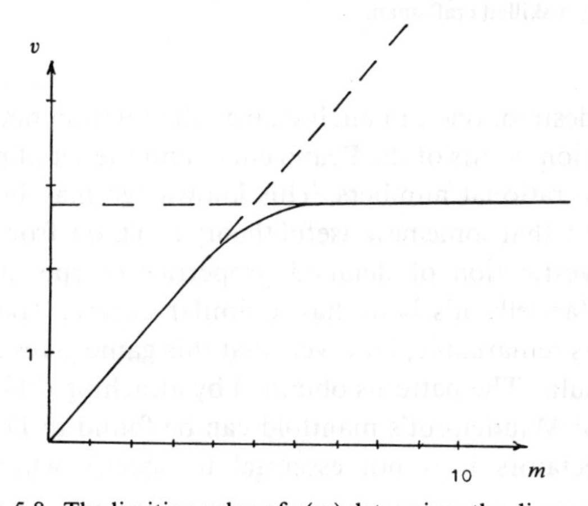

Fig. 5.8. The limiting value of $\nu(m)$ determines the dimension.

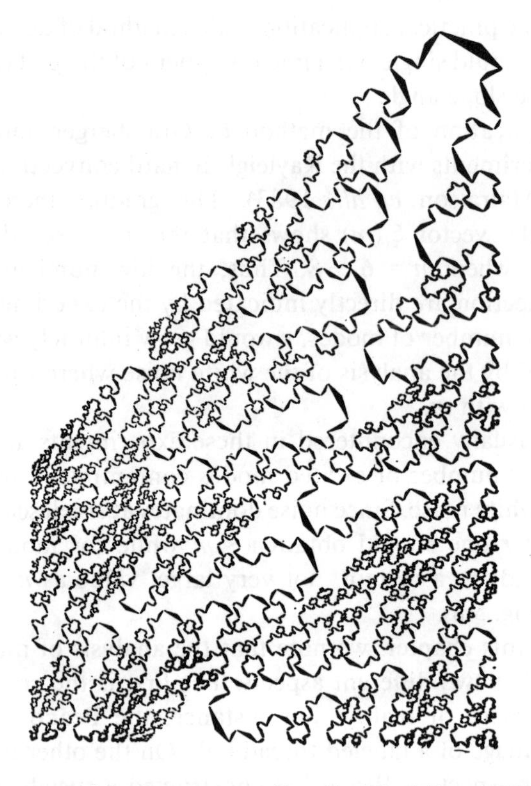

Fig. 5.9. The structure of the fractal set considered by Mandelbrot (1977) resembles a piece of fabric knitted by a skilled craftsman.

such curve. If desired, one can analyse, e.g., the relation between the set of self-intersection points of the Peano curve and the set of points whose coordinates are rational numbers. This knowledge may be useless but nobody can deny that some new useful theory can grow from one of such seeds. The investigation of detailed properties of specific manifolds described in Mandelbrot's book has a similar nature. Today this is a mind game. It is remarkable, however, that this game gives aesthetically satisfactory results. The patterns obtained by attaching different colours to the details of Mandelbrot's manifold can be found in Duday (1985). For many spectators it is not essential to specify which particular manifold was used by Mandelbrot for illustrating the properties of fractals and which methods were used for visualizing these patterns; it may be more interesting that these pictures would remind them of the colour fantasies of G. Klimt and the technique of automatic painting invented by the surrealists (see, e.g., Janicka, 1985); for others who have business in the textiles industry, these pictures may mean profits (see Fig. 5.9).

CHAPTER

SIX

PERCOLATION AND RANDOM BOUNDARIES

6.1. CLUSTERS AND PERCOLATION

In Chapter 5 we have mentioned a model of a ball lightning in which the active matter distribution forms a fractal cluster. The fractal properties of the matter distribution make the combination of a small weight (almost the whole volume of the ball lightning is free of active matter) and great strength (even though the active matter occupies a small fraction of volume, it penetrates to all parts of the lightning volume) possible. It is clear that a suitable fractal dimension of the cluster is not sufficient to reach the required strength of the construction. For a given fractal dimension, the matter would be concentrated in numerous relatively small and weakly connected clumps. The strength of the whole construction would be insufficient. Apart from the fractal dimension, the degree of internal connection plays an important role. It would be too restrictive to require that all particles of the active matter are in contact with one another (this is how the connection is understood in topology): it is admissible that there are a few pieces or clumps within a ball lightning. A more realistic picture is one that requires that there are connected parts whose sizes are of the order of the overall size of the ball lightning. Now we can distract ourselves from the fact that a ball lightning has a finite size and we consider the problem at scales much smaller than the lightning size. This brings us to an idealized description of the fractal cluster of active matter within a ball lightning: to ensure its strength, connected components that stretch from infinity to infinity must be present. Thus, we have approached an important concept of modern mathematics — percolation.

In our discussion, percolation appears to be connected to the model of a ball lightning. Of course, this is not the most practical application of the percolation theory. More important are the applications to the theory of compositional materials which consist of a mixture of ingredients with various properties. Consider the conductivity of a sample made of a pressed mixture of pieces of conductors and insulators. Clearly, the conductivity essentially depends very much on whether or not the electric current can flow along the system of conducting pieces. This phenomenon gives rise to the term "percolation" applied in various areas of science. Another important application of the percolation theory is in cosmology where the percolation properties of matter distribution in the universe are analyzed.

Mentioning the example of a composite material pressed from a metal and a dielectric, we imply that we have a sufficiently big, homogeneous piece in which the contents are oriented, on the average, isotropically and placed randomly. In a general case, the percolation theory usually considers the percolation properties of random sets in an infinite space when probability distributions, which determine these sets, are homogeneous and isotropic.

Nowadays, acceptable conceptions of the properties of percolation objects have been formed. In a popularized form, they are summarized by Efros (1982). First, it seems that the possibility of percolation depends only weakly on the shape and detailed structure of elementary objects forming a percolating cluster (pieces of metal in the example above). This hypothesis, known as the hypothesis of universality, is plausible as long as the size of the percolating cluster is much larger than the size of the individual pieces of metal. Next, it is clear that the percolation properties essentially depend on whether the considered object is one-dimensional or three-dimensional. In one-dimensional objects, percolation is practically impossible: any small particle of an ideal insulator disconnects the one-dimensional electrical circuit. In two dimensions, percolation over the metal phase implies impossibility of percolation over the insulator phase and vice versa, provided that the considered system is statistically isotropic. It is then natural to assume that in this case percolation is possible over the phase which occupies more than half of the area. In three dimensions, percolation can occur over both phases simultaneously. One can therefore expect that for small concentrations of the metal phase percolation over this phase is impossible, for higher concentration percolation occurs over both phases, and for still higher concentration of the metal phase percolation occurs over this phase exclusively. When the statistical properties of both the metal and dielectric phases depend identically on concentration, the thresholds for percolation (i.e., the value of concentration C_m for which percolation begins over the metal phase and the value C_d for which percolation ceases over the dielectric phase) are connected by an obvious relation

$$C_m = 1 - C_d\,.$$

Numerical simulation gives $C_m \approx 0.16$. This important figure will be frequently used below. Thus, as little as about twenty percent of the conductivity admixture make a composite material conductive. When the fraction of metal reaches approximately eighty percent, percolation over the dielectric becomes impossible; it becomes isolated into separated pieces.

The theory of percolation is now a vast field of mathematics and physics where the results are based on rigorous theorems as well as on numerical simulations and physical experiments. During the last thirty years, the pioneering work of English physicist and engineer Brodbent and mathematician Hammersley has given birth to an immense number of papers. In order to give a more concrete idea of applications of these results, we consider an example of the percolation properties of magnetic field lines (Zeldovich, 1983).

Importance of magnetic lines, which are determined by the equation

$$d\mathbf{r} = \mathbf{B}(\mathbf{r})\,d\alpha \tag{1}$$

(where α is an arbitrary scalar parameter), has been revived in recent decades, many years after their introduction by M. Faraday, mainly in connection with the problem of controlled thermonuclear fusion and problems of astrophysics, where the magnetic field is frozen into a plasma. In this case the plasma distinguishes a natural reference frame in three-dimensional space, comoving with the fluid, in which a (pseudo) vector $\mathbf{B}$ is determined (in contrast to the electromagnetic field tensor F_{ik} in empty Minkowski's space where there is no such frame).

Let us consider the topological properties of systems of magnetic lines. One of the important questions is: can magnetic lines extend to infinity in this or that direction? One can replace magnetic lines by narrow pipes filled with a liquid and ask whether or not a given system of pipes can transport the liquid to infinite distances. This percolational statement of the problem is complementary to the analysis of linkages of magnetic lines thoroughly developed by Moffatt and other authors (see Moffatt, 1969; Zeldovich *et al.*, 1983; Ruzmaikin and Sokoloff, 1980).

Percolation along magnetic lines is interesting mainly because the charged particles are spiralling along these lines. Evidently, the thermal conductivity of fusion plasmas, as determined by diffusion of electrons, depends on the properties of the magnetic lines and the surfaces around which the lines are wound. In particular, Kadomtsev and Pogutse (1979) consider a three-dimensional problem in which a weak random two-dimensional magnetic field $\mathbf{b} = (b_x, b_y, 0)$ is superimposed on a strong uniform field $\mathbf{B} = (0, 0, B_0)$ directed along the z-axis. Diffusion and heat conduction are determined by the tangling of magnetic lines associated with the presence of b. The two-dimensional field b can be expressed through a scalar function a of two variables, i.e., the z-component of the vector potential $\mathbf{a} = (0, 0, a)$:

$$b_x = \frac{\partial a}{\partial y}; \quad b_y = \frac{-\partial a}{\partial x}. \tag{2}$$

On the plane (x, y), the magnetic lines wound around the maxima in a counter-clockwise direction and around the minima in a clockwise direction.

When a is independent of both the time t and the z-coordinate, the problem reduces to the percolation properties of a random function of two variables.

Let us adopt the following normalization:

$$\langle a \rangle = 0; \quad \langle a^2 \rangle = 1. \tag{3}$$

Let us also consider the sufficiently smooth functions a, whose autocorrelation function is also sufficiently small at large distances. Using a spectral representation, this means that the Fourier transform of a has random phases and amplitudes a_k such that, e.g.,

$$\langle a_k^2 \rangle \sim \exp\left(-\frac{k^2}{k_0^2}\right), \quad k > k_0\,,$$

and (4)

$$\langle a_k^2 \rangle \sim \left(\frac{k}{k_0}\right)^{2n}, \quad n > 0, \quad k \ll k_0\,.$$

It is natural to expect that the regions where $a > \varepsilon$ (with $\varepsilon > 0$), which occupy less than half the total area, are isolated (islands) and there is no percolation over them. One may also introduce the quantity $l(\varepsilon)$ that characterizes the average size of an island or the average length of the isoline that surrounds an island.

In a two-dimensional problem, when the plane is divided into regions of two types, it is natural to assume that when the regions of one type form isolated islands, the regions of the other type form a globally connected ocean. Correspondingly, the conditions $a < \varepsilon$ and $\varepsilon > 0$ determine a unified region along which percolation occurs.[a] The value $\varepsilon = 0$ is critical with respect to the possibility of percolation.

Kadomtsev and Pogutse (1979) obtained an estimate of the function $l(\varepsilon)$ and suggested that the properly averaged value of $l(\varepsilon)$ plays the role of the effective free pathlength in the theory of diffusion and heat conduction.

In plasma devices with strong longitudinal field $B_z = B_0$, there is no reason to believe that the weak perturbations, **b**, are independent of z and/or time. Let us consider an opposite case when (i) either the field B_0 directed along z is absent or there is periodicity along the z-axis, with period $2\pi R$ that corresponds to a torus of large radius R with the z-axis chosen along the torus' large circumference; (ii) the two-dimensional field **b** is independent of both z and t; and (iii) the total field is strong and

[a] Some parts of the region where $a < \varepsilon$ can form lakes isolated from the percolating ocean.

charged particles move along the magnetic lines, i.e., the Larmor radius r of the electrons' spiral trajectories is neglected as well as the collisions and other effects that can cause jumps of the electrons from one magnetic line to another.

Due to this change of the magnetic field configuration, the solution of the considered problem becomes quite different from that obtained by Kadomtsev and Pogutse (1979): in a two-dimensional random field, diffusion of electrons is impossible, i.e., diffusivity is zero. To express this result in a constructive form, this means that the diffusion approximation is applicable only in such large space-time scales where the dependence on z and t is essential. At smaller space-time scales, the turbulent diffusion approximation is inapplicable.

This result is based on the fact that orbits of the electrons in the (x, y) plane turn out to be closed. When the initial smooth distributions of the electron number density n and the temperature T are given, the motion along the closed orbits leads to nothing else but only averaging over the orbit. In this approximation, the values averaged over the orbits, $\bar{n}$ and $\bar{T}$, remain always different for different orbits.

Another formulation of the problem considers a layer of finite thickness, e.g., $0 < x < x_0$, where the considered two-dimensional random magnetic field is concentrated. The currents flowing along the z-direction that produce the field b are absent on both sides of this layer.

Let us prescribe the value n_1 of the number density to the left of the layer $(x < 0)$ and another value n_2 to the right $(x > x_0)$. The particle flux is given by

$$q_x = D(n_1 - n_2)/x_0 , \tag{5}$$

following the definition of coefficient of diffusion D. It is now clear that with the growth of the layer thickness x_0 the fraction of orbits which are not closed within the layer (i.e., between $x = 0$ and $x = x_0$) decreases. Therefore, the particle flux decreases more rapidly than x_0^{-1}, either as x_0^{-m} with $m > 1$ or as exp $(-k_0 x_0)$. But this implies that there is no definite value of D and in the limit of large x_0, the effective value of D tends to zero.

A finite (non-vanishing) value of D can be obtained only when the Larmor radius r is finite.

In a similar problem of a two-dimensional steady vortex motion of incompressible fluid, the turbulent diffusivity is non-vanishing only when the molecular diffusivity k is non-vanishing (Zeldovich, 1982). In this case D is proportional to a certain fractional power of k.

In this respect, a two-dimensional steady motion is similar to a one-dimensional unsteady motion (Zeldovich, 1982). The analogous correspondence with the patterns of caustic lines (or surfaces) in problems with equal numbers of variables (e.g., either x and t or x and y) was noted by Arnold (1982).

In similar completely three-dimensional stationary problems as well as in two-dimensional non-stationary problems (three variables, either x, y, z or x, y, t), both the electron diffusivity D along the magnetic lines and the turbulent diffusivity D_t of hydrodynamic motion differ from zero even in the limits $r \to 0$ and $k \to 0$.

To verify this, consider, e.g., a non-stationary two-dimensional percolation problem for a system of magnetic lines with arbitrary unsmooth dependence of the magnetic field on time. Here a particle moves along a closed field line around some center during the time interval τ_1 and during this period its coordinates differ from their initial values by at most the orbit radius. In the following time interval τ_2, after the magnetic field has abruptly changed, the particle can jump to another center and move along a new trajectory which is independent of the previous one.

Evidently, after a few such steps, the displacement grows with the square root of time, that is, according to the typical diffusive law.

Let us consider briefly the percolation properties of a two-dimensional field consisting of two components, a steady uniform field and a randomly varying field. The weak random field bends the field lines of the uniform field only slightly. It is clear that the only physical effect is the following. In some places, where the random field is sufficiently strong, local catastrophes lead to the isolation of islands of closed streamlines associated with the local maxima of a accompanied by saddle points.

More interesting is the opposite case when a very weak uniform steady field is imposed on a given random field. It can be shown that the magnetic flux corresponding to the uniform field concentrates into narrow ropes (channels) along which occurs percolation in the direction of the uniform field, even though the ropes are tangled. Within the ropes the field strength is of the order of the r.m.s. chaotic field; weakness of the uniform field leads to the narrowness of the ropes.

Note one more particular case. Consider a uniform field in an ideally conducting medium. The random motion of the medium amplifies the field. The flux is conserved because under chaotic tangling the mean cosine of the angle between the current field direction and the original one tends to zero. The percolation remains to be ideal until the field is frozen into the medium and the field lines do not reconnect. The pattern of narrow percolation that channels here arises only after a sufficient time has elapsed from the beginning of the motion.

To conclude this discussion of percolation along the magnetic lines of a two-dimensional field, we should emphasize that although this picture seems to be quite natural, it implicitly relies on a non-trivial restriction on the magnetic field configuration. This restriction can be conveniently expressed in terms of the correlation properties of electric currents that produce the considered magnetic fields; recall that in the two-dimensional case, the current j is directed along the z-axis and is related to the potential a through

$$\Delta a = j\,.$$

It can be shown (Zeldovich, 1983) that the discussion above implicitly assumes that the electric currents are positively correlated, i.e.,

$$\langle j(\mathbf{x})\, j(\mathbf{x} + \mathbf{r})\rangle \geq 0\,. \tag{6}$$

In particular, the discussion above is applicable when the considered magnetic field is produced by currents that are directed completely at random (are uncorrelated). However, the situation can be very different when the currents are screened and thus compensated. Percolation in a more complicated system is considered in the following section where the cosmological percolation problem serves as an example.

6.2. INTERMITTENCY AND PERCOLATION

The picture described in the previous section reflects only one aspect of percolation even though it is very important. In order to reveal another aspect, let us turn to matter percolation in the universe. Presently, the average matter density in the universe is much below the density of any cosmic body, from planets to galaxies and galaxy clusters. This implies that matter is distributed uniformly only over a large scale while at smaller scales the distribution is highly non-uniform: vast voids coexist with compact clumps of matter. The volume fraction occupied by these clumps (specifically, the galaxy clusters) is considerably smaller than the constant of percolation, 0.16. Thus, the results of Section 6.1 seem to imply that galaxy clusters should form isolated islands separated by voids. Such distribution of matter in the universe coincides with the clustering model which assumes that the observed structure of the universe is the result of progressive hierarchical clustering of matter at much larger scales. It is noticeable that a careful analysis of observations definitely rejects this model (see the review of Zeldovich and Shandarin, 1982) and it can be consistent with observations only with a very low probability. In other words, the galaxy clusters cannot be described as balls widely scattered according to the Poisson law. In order to reach a correct, rather than apparent, conclusion about matter distribution in the universe, one should remember (Zeldovich, 1983) that at the earlier stages of evolution of the expanding universe matter distribution was quite different: almost all the matter was in the state of high density, exceeding the average density of the present universe. At these epochs, the matter that now forms clusters of galaxies occupied a major fraction of the universe volume. The results of Section 6.1 then imply that percolation at that time must proceed along these dense regions while the future voids must be isolated. The present low-density state of the universe is produced from a dense state of the early universe due to expansion which can be considered here as continuous mapping. Obviously, a continuous mapping cannot affect the percolation properties of the objects, i.e., even now the percolation must occur along the system of galaxy clusters.

Thus, we have reached a paradoxial result: the concepts of percolation discussed in Section 6.1 seem to imply that both clusters and voids must be isolated in the present universe.

The paradox is solved as follows. It seems plausible that matter inhomogeneities in the early universe are approximately spherical in shape and the percolating structure of the dense regions is formed by contacts of these spherical regions. The situation is quite different at the present epoch: during the expansion of the universe, the gravitational instability, to the first approximation, compresses the matter inhomogeneities along one direction only while expansion still can occur along the remaining two directions (Zeldovich, 1970). As a result, matter is concentrated within flat, thin formations — like pancakes. Thus, the percolation theory for the present universe should incorporate an additional characteristic parameter, a small ratio of the thickness of the pancake to its diameter. The value of this parameter is considerably smaller than the value 0.16 of the three-dimensional percolation parameter; percolation properties of the matter distribution are determined by the former parameter. Gravitational instability leads to the pronounced cellular structure of the matter distribution in which large voids are separated by very thin walls that contain the principal part of the total mass.

We see that the intermittency in the distribution of a random field can qualitatively modify the percolation properties. Percolation properties of a system of thin, long strings or plates can considerably differ from those of a system with more or less spherical bodies considered in Section 6.1. This kind of percolation is very important for many applications. For example, the implantation of flat flakes of suitable impurity into a polymer can lead to a considerable slowing down of its deterioration with time. Modern advances in the theory of percolation in intermittent media are reviewed by Menshikov *et al.* (1986). The authors are much indebted to S.A. Molchanov for numerous discussions of the percolation theory reflected in this chapter.

In three-dimensional intermittent systems, percolation seems to be associated with the presence of a rather long and thin structure. For moderate ratios of the impurity bodies or for low concentrations, their role seems to be negligible. An implicit indication of this is a low value of the percolation threshold even for spherical particles, $C_m \approx 0.16$.

The situation is different in two dimensions where percolation is not connected with any small parameters. In particular, Menshikov *et al.* (1986) proposed the following picture of two-dimensional intermittent percolation: for small concentrations of the conductor phase, percolation cannot occur along it, for its large concentration stops percolation over the insulator phase. However, unlike non-intermittent distributions, for intermittent distributions transition between these two regimes occurs before the conductor phase reaches half the volume. Around the state with equal concentrations of both phases, there is a finite interval of concentrations for which percolation does not occur along either phase. The structure in these cases is similar to a set of nested alternating closed layers of conductor and insulator. What occurs is a local spontaneous violation of translational symmetry and the matter distribution resembles a polycrystal whose individual crystals consist of such nested layers. An isolated set of nested layers has a distinguished center. A nested-layer monocrystal has finite size but a large number of such monocrystals fill the pores of a larger nested-layer monocrystal. The resulting hierarchy of nested-layer monocrystals is translationally invariant (see Fig. 6.1). In finite bodies, such a structure leads to irregular changes of conductive to insulating properties under weak changes of size of samples.

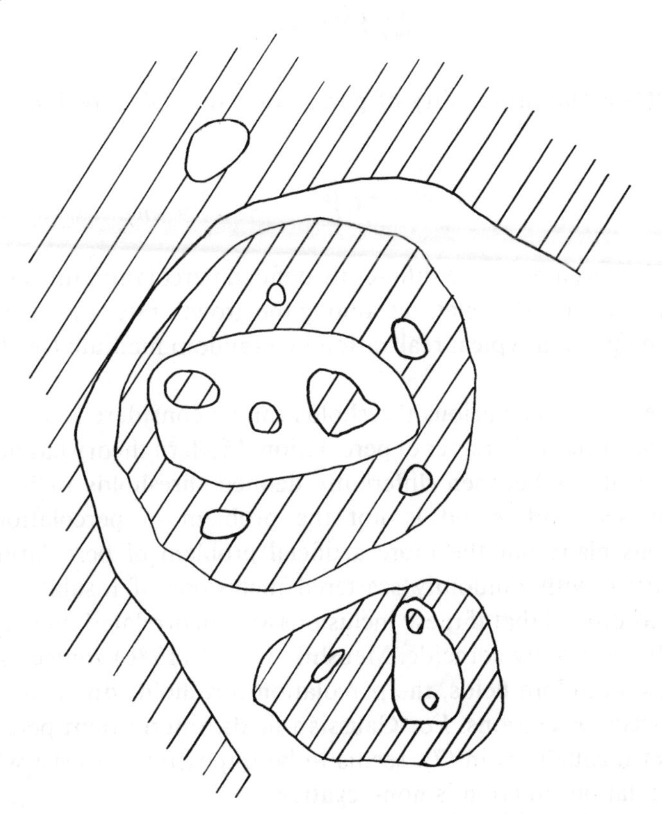

Fig. 6.1. Qualitative scheme of nested structure for which percolation is impossible over either phase in two dimensions.

These ideas can be expressed also in another way. For intermittent media, the standard ergodic concepts are typically violated. In the considered case, intermittency leads to ambiguity of the notion "percolation threshold". This threshold can be understood as a critical value of the concentration at which exists an infinite cluster of a given phase (C_H), or as the value of concentration at which the mathematical expectation value of the cluster size becomes infinite ($C_T^{(1)}$). One can also introduce an intermediate percolation threshold, $C_T^{(p)}$ for which the p-th moment of the cluster size becomes infinite. Finally, one more quantitative characteristic of the percolation can be introduced, that is the value of the concentration C_s for which

$$\lim_{l \to \infty} P(l) = 0 ,$$

where $P(l)$ is the probability of percolation through a region of size l. Clearly,

$$C_s \leqslant \ldots \leqslant C_T^{(p+1)} \leqslant C_T^{(p)} \ldots \leqslant C_T^{(1)} \leqslant C_H . \tag{7}$$

These relations among the differently defined percolation thresholds can be compared with the relations among the growth rates of the magnetic field moments in a typical realization in a random medium (see Chapter 9).

Differences in the various thresholds can be considered as a criterion of the intermittent character of percolation. Modern theory has not as yet studied relations between differently defined thresholds in full detail. More or less understood is not the problem of percolation on a continuous plane but the more artificial problem of percolation on a metal lattice with randomly scattered inclusions of insulator. Kesten (1982) has proved that if these inclusions are uncorrelated, then all kinds of thresholds exactly coincide. Menshikov *et al.* (1986) conjectured that for Poisson random fields, the percolation thresholds differ due to well pronounced correlations. For Gaussian fields, intermittent percolation, i.e., strict inequalities in (7), seems to be impossible, at least when the field correlation function is non-negative.

In its initial formulation, the percolation theory considers percolation along purely random regions. This theory is related to another realm of problems in which the presence of some degree of regularity is essential, e.g., in the case of regions with regular but randomly deformed boundaries. Such regions are exemplified by deformed or imperfect resonators. The theory of such resonators has some common aspects with the theory of percolation, and we shall briefly consider them in the following section (Zeldovich and Sokoloff, 1983).

6.3. EIGEN-OSCILLATIONS OF A REGION WITH RANDOM BOUNDARY

Modern mathematical physics considers the analysis of generic problems (see, e.g., Arnold, 1983) very important. These problems often drastically differ from exactly solvable problems where solvability is associated with a high degree of symmetry. One particular case of a generic problem is the problem with statistically (randomly) determined parameters.

Here we consider the question of eigen-oscillations of a region with randomly determined boundary, i.e., the question of eigenfunctions of the problem

$$\Delta u = -\lambda^2 u, \quad u_{|\partial\mathcal{G}(\omega)} = 0,$$

where $\partial\mathcal{G}(\omega)$ is the random boundary and ω is the random parameter (Babich and Buldyrev, 1972).

First consider a one-dimensional problem where the random region is a line segment $[a(\omega), b(\omega)]$. Although the solutions to this problem are random functions, the n-th eigenfunction has, like the deterministic one, $n - 1$ zeros in the interval (a, b) so that the distance between the zeros is of the order of $(b - a)/n$. In the one-dimensional case, the requirement of generic properties does not introduce novel properties into the structure of the zeros of an eigenfunction. The situation is more complicated in multi-dimensional cases.

Consider now a two-dimensional region $\mathcal{G}(\omega)$ whose boundary is close to a regular curve of diameter $2R$, e.g., to a circle of radius R or a square of side R. This regular curve is disturbed by a random function which is locally similar to the Wiener process, i.e., $\Delta\xi \sim (\Delta l)^{1/2}$, where $\Delta\xi$ is the deviation from the unperturbed curve and Δl is the coordinate measured along the regular curve.

The eigenfunctions have zero lines within the region $\mathscr{G}(\omega)$, which are analogous to the zeros of one-dimensional eigenfunctions. The zero line patterns can be used for topological classification of the eigenfunctions. Excitation of the next higher oscillation mode adds an extra zero line. Due to the random nature of the boundary, this new line with probability unity does not intersect other zero lines. Indeed, level lines $u = 0$ of a random function $u(r)$ have no singular points. The intersection of two zero-level contour lines is not a generic event in the random media, i.e., in the space of the surfaces the measure corresponding to this event is zero. A formal proof of this fact can be found in Brüning (1978). Therefore, n zero lines divide the region $\mathscr{G}(\omega)$ into $n + 1$ parts. These zero lines must either be close to or intersect the region boundary. Of course, in contrast to the one-dimensional case, the number of zero lines now cannot completely describe an eigenfunction; an accurate classification should be taken into account as well as the topology of the network of zero lines.

We should stress that the absence of any symmetry in the shape of the considered region implies that the distance between zero lines is of the order of R/n, i.e., zero lines can approach each other closely only with a small probability or for large n.

When there is some symmetry, the set of zero lines can have a more complicated structure. Due to symmetry, zero lines can have numerous mutual intersections so that $m + n$ zero lines parallel to the sides of a regular square divide the square into $(m + 1)$ $(n + 1)$ regions (see Fig. 6.2). R. Courant (see Hilbert and Courant, 1981) has obtained an estimate of the number of such regions as a function of the eigenfunction's order. He has also noted that in degenerate situations, including the problem of oscillations in a square, only a few very symmetric eigenfunctions provide that large number, $(m + 1)$ $(n + 1)$, of regions separated by zero lines. A linear combination of eigenfunctions with generic coefficients behaves similarly to an eigenfunction of a statistical problem. The intersection points of zero lines are destroyed while the zero lines become closed. However, in this case the intersections of zero

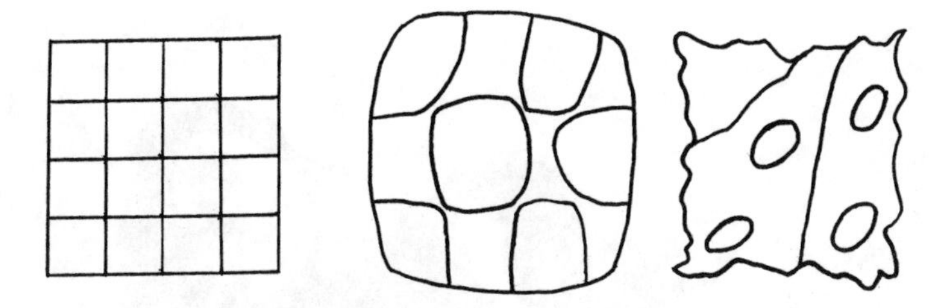

Fig. 6.2. Schematic structure of zero lines of eigenfunctions in problems (a) with symmetry; (b) with weak deviation of symmetry due to boundary perturbation; (c) with strongly violated symmetry.

lines are destroyed due to violation of the symmetry by the coefficients of the linear combination of symmetric eigenfunctions, rather than due to violation of the region's symmetry.

There exists a class of perturbations of a symmetric region that leads to, in a certain sense, weak perturbation of the pattern of zero lines even though the intersection points are destroyed (see, e.g., Babich and Buldyrev, 1972; Lazutkin, 1979). It turns out that when a circle is perturbed in such a way that the curve remains convex, the distance between neighbouring closed zero lines is exponentially small in the perturbation, rather than of the order of R/n. In this case, perturbation of the boundary is too weak to lead to a considerable change of the zero lines pattern.

One can also analyze the opposite case of very strong perturbations which drastically change the picture. This change is associated with the fact that, apart from eigen-oscillations localized far from the boundary, which were considered above, there arise in a region with random boundary special boundary-layer eigen-oscillations of the type of Echo Wall.[b] Their existence is associated with the irregularities of the boundary and interference of the eigenwaves at these irregularities. Such

[b] The Echo Wall, known as the Whispering Galleries in Russian, a masterpiece of Chinese architecture in Beijing, more than adequately describes the phenomenon. A whispering sound propagates very far along these walls, to distances where one cannot normally hear a loud voice originating from the same location (Fig. 6.3).

Fig. 6.3. The Echo Wall in Beijing always attract tourists (the illustration from Peng Zhen, 1986).

oscillations decay exponentially with distance from the boundary and the question of the structure of zero lines is of minor importance for them. However, if the surface diameter $\sim 2R$ is much greater than the perturbation amplitude and, in addition, the perturbed boundary length does not differ by an order of magnitude from $2R$, then "whispering" perturbations are negligibly deep within the considered region. The situation is different in three (and more) dimensions. In such a case, the boundary area ($\sim R^2$) grows faster than its diameter and the large area of the boundary leads to the plausible appearance of high peaks of height $\sim R$ even if the probability of such a peak is low. Therefore, in the three-dimensional case, the dependence required for the damping of boundary-layer perturbations is $\Delta\xi \sim (\Delta l)^{3/2}$.

Let us now discuss how these problems are connected with the percolation theory. To do this, one should consider a double limit case when both the region size and order of the eigenfunction simultaneously tend to infinity so that the characteristic value of the wave number constant. In this case, the problem reduces to that of a random function φ with zero mean value whose spectrum is concentrated in the "sphere" $|K| = K_0$ (see Section 6.1).

Now we are able to formulate the percolation problem, i.e., the question whether or not it is possible to pass from infinity to infinity, e.g., along the regions where the eigenfunction φ is positive. In three dimensions, the possibility of percolation is typical. For a very large ε, exceeding some critical value ε_0, percolation is impossible along the region with $\varphi > \varepsilon$; for $\varepsilon_0 > \varepsilon > -\varepsilon_0$ percolation occurs over both regions with $\varphi > \varepsilon$ and $\varphi < \varepsilon$; for smaller ε percolation along the region with $\varphi < \varepsilon$ is impossible. Roughly, the value of ε_0 is of the order of $\langle \varphi^2 \rangle^{1/2}$ which means that for a Gaussian random variable the region with $\varphi > \varepsilon_0$ occupies 1/6 of the total volume. This problem is much more complicated in two dimensions where simultaneous percolation along both phases is impossible. In the two-dimensional case, the simplest possibility is the following. For any $\varepsilon > 0$ the lines $\varphi = \varepsilon$ and $\varphi = -\varepsilon$ delineate a system of islands with positive and negative values of φ. There can exist a system of lakes with φ of opposite sign within any island. When ε tends to zero, the picture remains qualitatively the same: the sea between the islands reduces to a system of channels but the islands remain isolated. However, one can envisage another intermittent picture: at a certain value of ε a "phase transition" occurs and the system of islands and channels is replaced by a nested-layer structure with islands of smaller size nested in larger islands with opposite sign of φ so that the whole island hierarchy is infinite.

6.4. THE ZEROS OF EIGENFUNCTIONS OF FREE AND FORCED OSCILLATIONS

The question of the structure of the manifold in which the oscillation amplitude in a resonator vanishes is also close to the percolation problem. In the simplest case, for a three-dimensional resonator this manifold consists of surfaces. However, in the generic case the oscillation amplitude vanishes at one-dimensional manifold lines which are closed or otherwise begin and end at the region boundary. After traversing along a closed zero-amplitude line, the oscillation phase changes by $\pm 2\pi$.

The appearance of topological singularities of filamentary structure whose traversal changes the phase by $\pm 2\pi$ is typical of many physical problems. Polyakov (1975) has noticed such singularities in the theory of spontaneous symmetry breaking. Consider a complex scalar field ψ. For the fields that appear in the cosmological inflation theory, the vacuum corresponds to $\psi = \psi_0 \neq 0$. If the phase changes by $\pm 2\pi$ after traversing along a closed contour, this contour must enclose a line (string) at which $\psi = 0$. Similar ideas can be developed for auto-oscillatory systems. Aldushin *et al.* (1980) consider the surface of a flame in the case where its propagation at constant speed is unstable, while nonlinear effects stabilize oscillations of the flame front at a certain amplitude and phase and the transverse dissipation equalizes the phase over the whole flame surface. If the initial conditions are such that at a certain contour on the flame surface the phase changes by $\pm 2\pi$, then this contour must enclose a point with a vanishing oscillation amplitude. This singularity occurs at a point rather than a line because the flame surface is two-dimensional. Singular lines are typical of three-dimensional auto-oscillatory systems. Similar problems appear in the theory of chemical auto-oscillatory systems and in the theory of nearly parallel rays (Berry and Nye, 1974; Ivanitsky *et al.*, 1978; Baranova and B. Zeldovich, 1981).

Similar lines also arise in the simple case of a linear oscillatory system with periodic external forcing or dissipation.

To verify this, recall that the equation of oscillations supplemented by the boundary condition $u = 0$ at a closed surface represents a self-adjoint problem for the equation $\Delta u = -\omega^2 u$. Therefore, its solutions are real. Even if some eigenfunctions u_n are complex, the conjugated functions $\overline{u}_n$ are also solutions corresponding to the same eigenvalue ω; these pairs of eigenfunctions can be represented as pairs of real functions whose zeros lie on a surface. A similar result applies to the case of phased forcing, i.e., when the condition $u(s, t) = R(s) \exp(i\omega t)$ with real R applies to some part of the boundary surface.

Let us now turn to the case where zero surfaces are replaced by zero lines. The case of forced oscillations with unphased excitation,

$$R(s) = R_1(s) + iR_2(s), \quad \frac{R_1}{R_2} \neq \text{const.} \tag{8}$$

is simple and clear. Indeed, due to the linearity of the problem solutions have the form

$$u(t, \mathbf{r}) = \exp(i\omega t)\,[\rho_1(\mathbf{r}) + i\rho_2(\mathbf{r})]\,,$$

where ρ_1 and ρ_2 correspond to R_1 and R_2, respectively. Physical solution is given by the real part of this expression, $\mathrm{Re}\,u = \rho_1 \cos \omega t + \rho_2 \sin \omega t$, and the condition $u = 0$ leads now to two conditions, $\rho_1 = 0$ and $\rho_2 = 0$. Thus, the zero manifolds are now intersections of two surfaces, i.e., the lines. Note that for $R_1 \propto R_2$ we have $\rho_1 \propto \rho_2$ and $\mathrm{Re}\,u \propto \cos(\omega t + \varphi)$ with φ constant. Thus, in this case the zero manifold is a surface, and it reduces to a line only for $R_1/R_2 \neq \mathrm{const}$.

The situation is the same for oscillatory systems with damping or radiation. The oscillations are now described by the equation $u_{tt} = \hat{L}u$ where $\hat{L}$ is the differential operator. Consider a boundary-value problem for this equation which is not self-adjoint, e.g., due to specific boundary conditions. Evidently, in a generic case the eigenfunctions are complex-valued. Therefore, the amplitude of oscillations turns to zero on a line or, in the case of an arbitrary number of dimensions, on a manifold with co-dimension 2 (co-dimension is the difference between dimensions of the space and of the embedded surface).

V.I. Arnold has drawn our attention to the fact that the problem of zero manifolds of eigen-oscillations was discussed by mathematicians, in particular, problems like the oscillations of two-dimensional surfaces in four-dimensional space. However, this direction has not been developed too far. The reason for this is probably associated with the fact that in classical problems of mathematical physics, even in the presence of non-self-adjoint operators, the situation usually is not generic and complex-valued frequencies correspond to real-valued eigenfunctions.

Let us illustrate the situation with examples. Consider dissipation described by the term au_t. Then the boundary value problem for the oscillations is formulated as

$$(-\omega^2 + ia\omega)u = \Delta u\,, \quad \frac{u}{\Gamma}\Big|_b = 0$$

and, for constant a, the substitution $\lambda^2 = \omega^2 - ia\omega$ reduces the problem to one of undamped oscillations. Based on the linearized Navier-Stokes equation, let us now describe dissipation through the term $\nu(\Delta u)_t$. Then the corresponding boundary-value problem is given by

$$-\omega^2 u = \Delta u + i\omega\nu\,\Delta u; \quad \frac{u}{\Gamma}\Big|_b = 0\,.$$

For constant ν the substitution $\lambda^2 = \omega^2(1 + i\omega\nu)^{-1}$ again reduces the problem to a dissipationless one. Finally, consider a parabolic equation $u_t = (\nu + i\Omega)\Delta u$. For $u \propto \exp(\gamma t)$, the substitution $\lambda = \gamma/(\nu + i\Omega)$ again makes the problem self-adjoint in a homogeneous and isotropic case. In these cases the complex-valued solutions arise only when, e.g., the dissipation is inhomogeneous [cf. condition $R_1/R_2 \neq$ const. in (8)]. The zero lines also arise when the excited waves leave the considered region through a hole.

Consider now another case of waves in an infinite medium:

$$u(t, \mathbf{r}) = r^{-1}\exp(i\omega t - ikr)\,.$$

Due to planar symmetry, this complex solution differs from zero everywhere. However, for the superposition of a spherically symmetric spreading wave and a travelling planar wave (which introduces a distinguished direction at infinity):

$$u(t, \mathbf{r}) = [Ar^{-1}\exp(-ikr) + B\exp(ikx)]\exp(i\omega t)\,,$$

we see that the amplitude is zero when the two conditions are fulfilled: first, $r = r_0 = A/B$ and second, on the sphere $|\mathbf{r}| = r_0$ the phases $r - x = 2\pi n/k$ with $n = 0, 1, 2, \ldots$ coincide up to 2π. The zero lines are of a few (with the number dependent on the value of kr_0) parallels on the spheres determined by the wave front. This group of problems is important in the studies of weakly divergent light beams (Baranova and Zeldovich, 1986).

All these results rely heavily on the monochromatic nature of oscillations. Qualitatively new physical effects can be associated with quasi-monochromatic oscillations. An important example is provided by the Langmuir plasma oscillations whose wavelength is much greater than the Debye wavelength. As a result, there arises a natural distinction between fast and slow oscillations and the physical picture may reduce to a slow drift of the zero lines of fast oscillations.

6.5. THE MATHEMATICAL LANGUAGE OF THE PERCOLATION THEORY

Traditionally, mathematics is considered as a science of quantitative relations and spatial patterns. These two realms of mathematics were not always in a harmonious combination and relation. In the second half of the 19th century, it seemed that The Quantity had won this competition. Treatises on geometry had got rid of drawings and figures, geometrical problems had been translated into the language of equations and methods of mathematical physics had merged with the theory of differential equations. The Form won back in early 20th century when topology appeared. In this non-quantitative science, even an invariant is not necessarily a number. The modern development of mathematics can be compared to a symphony with two dominating colliding themes. One theme is the quantitative mathematics which becomes more and more computerized, pays less attention to the rigorousness of proofs and approaches further the applied sciences and engineering. Another theme is the non-quantitative mathematics — geometry, topology, etc. This branch of mathematics has found many unexpected common points with the humanities and arts. It is still developing adequate language of self-expression and is, therefore, much concerned with the rigorousness of its proofs. However, this branch also makes its first steps toward computers, declaring that the numerical nature of computer science represents only the embryonic state of development of this science. Applications of non-quantitative mathematics to physics and other natural sciences are still not numerous even though their number grows steadily. In this respect, the mathematical method of percolation theory is unique, being almost completely based on the achievements of non-quantitative mathematics. Unfortunately, it is a tradition of this field of thought to express the results in a very abstract way paying much attention to rigorous proofs. As a result, books on percolation theory written by mathematicians often

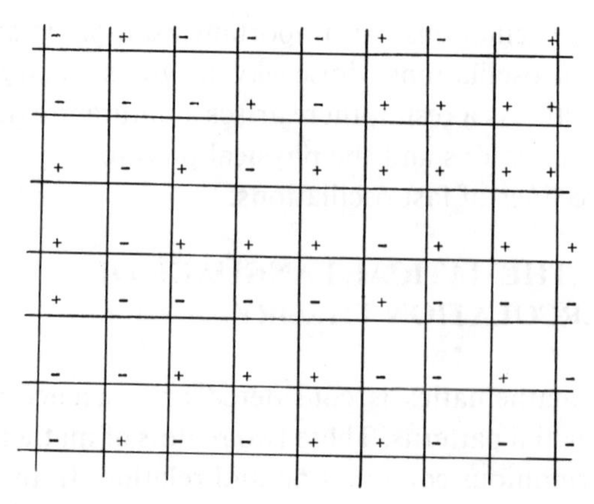

Fig. 6.4. In the knots problem, the knots conduct electric current with probability p (plus signs) and are insulating with probability $q = 1 - p$ (minus signs).

can only be understood by their authors. Percolation mathematicians are more than suspicious about the results obtained with the help of computers and persistently argue that simple and clear results obtained in percolation theory by physicists may prove to be inaccurate, having neglected a whole world of complex percolation phenomena like, e.g., intermittency. Hopefully, a synthesis of physical and mathematical approaches to percolation theory will be reached in the foreseeable future. But now we can only reflect this complicated state of lack of mutual understanding and give only a hint on specific methods of percolation theory.

The first step that mathematicians take in the studies of percolation is the discretization of space. Instead of percolation along a region in which a random field exceeds a given value, one considers percolation along a regular lattice with certain links removed with a given probability. It turns out that the result strongly depends on the manner of discretization, i.e., on whether one considers a regular square, triangular, hexagonal or more complicated lattices. The links also can be removed in different ways. For example, one can leave a chain crossing unaffected with probability p and remove it with probability $q = 1 - p$. This discretized problem is called the knot problem (see Fig. 6.4). Otherwise, one can leave a link unaffected with probability p and remove it with probability $q = 1 - p$ (Fig. 6.5). This is the link problem. These two

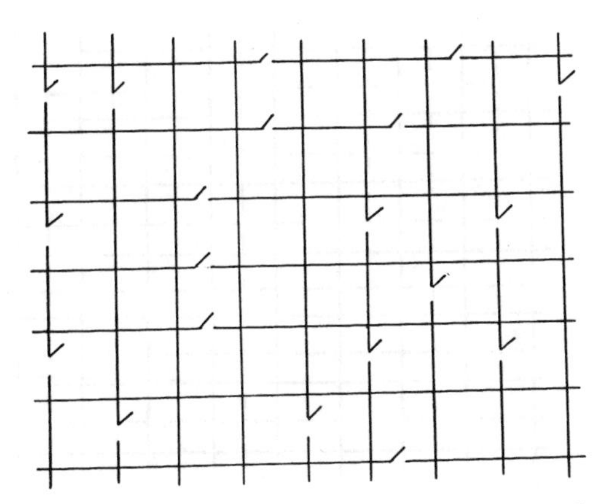

Fig. 6.5. In the links problem, the square lattice (solid lines) and the dual one (dashed lines) are shifted by a half-period with respect to each other along both axes.

kinds of formulation are seemingly analogous and one would expect that percolation thresholds (i.e., those critical values p_{cr} that correspond to marginal percolation) are close to each other. However, the actual link problem possesses higher symmetry than the knot problem. It turns out that the destructive role of a removed crossing is more pronounced than that of a removed link. To verify this, one should formulate the link problem in a symmetric form. This is achieved by the construction of the so-called dual lattice (Fig. 6.6) which is composed of connections between the cell centers of the initial lattice. A crossing of the dual lattice is considered present (switched on) when it crosses a removed crossing of the initial lattice and vice versa. It can be seen easily that if percolation occurs from the left to the right along the switched-on links of the initial lattice, then removed links of the dual lattice are percolating from above to below and vice versa. The absence of percolation along the switched-on links of the initial lattice implies percolation along the removed links of the dual one. Unfortunately, it is very difficult to turn this intuitive knowledge into a rigorously proved result (see Kesten, 1986). The existence of such symmetry implies that percolation along the switched-on links is immediately followed by percolation along the removed links when the critical value of probability is passed (i.e., the intermittent percolation is impossible). Since the properties of the initial and dual lattices are symmetric, this means that

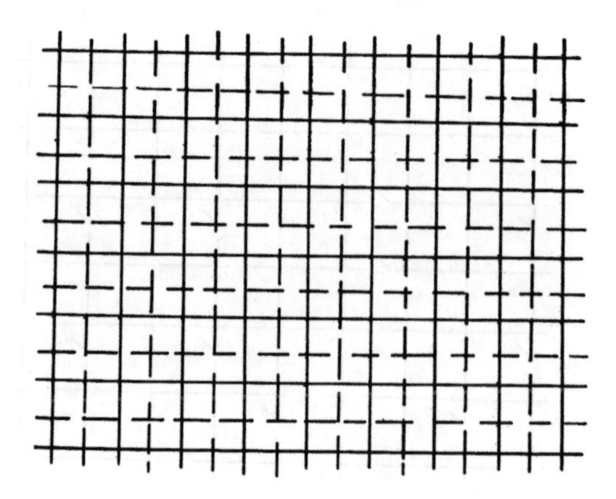

Fig. 6.6. A dual lattice composed of connections between the cell centers of quadratic lattice.

$$p_{cr} = 0.5 \,.$$

There is no such symmetry for the knot problem and a similar procedure brings us to a dual lattice whose percolation properties differ from those of the initial one. For this reason the percolation threshold exceeds one half in the knot problem. Numerical experiments give the value

$$p_{cr} = 0.59 \,.$$

Those cases where percolation thresholds can be estimated at the mathematical level of rigor, without the aid of computers, are also based on symmetry arguments, even though they are much more complicated and subtle than in the problem considered above. Sometimes the problem can be reduced to some equations. For instance, symmetry arguments lead to the conclusion (see Efros, 1982, pp. 171–172) that the percolation thresholds for the link problem on a triangular lattice obeys the equation

$$p_{cr}{}^3 - 3p_{cr} + 1 = 0 \,.$$

Verification of the fact that this equation has only one root

$$p_{cr} = 2 \sin \frac{\pi}{18}$$

belonging to the interval $0 < P_{cr} < 1$ or an approximate estimation of this root is a problem of quantitative mathematics, and the estimation can be obtained easily with the help of a computer. However, these computational problems are incomparably easier than those that arise in an attempt at direct numerical estimation of the percolation threshold.

CHAPTER

SEVEN

RANDOM HYDRODYNAMIC MOTIONS

A wide range of random phenomena is associated with hydrodynamic motions. The two principal problems of modern hydrodynamics — transition from a laminar flow to the turbulent one and the development of coherent structures in turbulence — are connected with the interplay of chance and order. As routinely emphasized in monograph studies and reviews devoted to turbulence, the agent that destroys one of the regimes or induces transition to another one is the nonlinear interactions of the normal modes of the flow. This kind of approach considers randomness either as a result (in transition from a laminar flow to the turbulent one) or as a statistically prescribed background (developed turbulence) against which the evolution proceeds. However, the role of chance in hydrodynamics deserves to be considered more carefully and discussed in more detail. Leaving aside attempts to give a complete description, we demonstrate here some selected aspects and intricacies associated with random hydrodynamic motions. To begin, we shall consider a kind of hidden turbulence.

7.1. THE TURBULENT FLOW WHOSE VELOCITY IS LAMINAR

Consider the flow that has a stationary, smooth, 2π-periodic, solenoidal velocity field:

$$v_x = A \sin z + C \cos y\,,$$

$$v_y = B \sin x + A \cos z\,, \tag{1}$$

$$v_z = C \sin y + B \cos x \,,$$

where A, B and C are real constants. As shown by Arnold (1965), this flow obeys the stationary Euler equation

$$(\nabla \times \mathbf{v}) \times \mathbf{v} = \nabla\left(\frac{p}{\rho} + \frac{v^2}{2}\right),$$

with the obvious property that both the vectors $\mathbf{v}$ (the streamlines) and $\nabla \times \mathbf{v}$ lie on the surfaces of constant enthalpy $p/\rho + v^2/2$ (because their cross-product is directed orthogonally to these surfaces).

At first sight, the field (1) has nothing particularly interesting except the fact that it coincides with its curl, $\nabla \times \mathbf{v} = \mathbf{v}$, i.e., belonging to the class of Beltrami flows. In other words, the vector field (1) possesses helicity, $\mathbf{v} \cdot (\nabla \times \mathbf{v}) \neq 0$, and in this flow helicity attains its maximum possible value.

Surprising things emerge when one considers trajectories of the fluid particles in the velocity field (1), which are governed by the equations

$$\frac{dx}{v_x} = \frac{dy}{v_y} = \frac{dz}{v_z} \,.$$

Performing numerical integration of these three equations for the case $A = 3^{1/2}$, $B = 2^{1/2}$ and $C = 1$, Hénon (1966) discovered regions of finite sizes within which the trajectories are chaotic while they are quite regular in the surrounding areas. A detailed study conducted later by Dombre *et al* (1986) has shown that this picture is typical of all non-vanishing values of A, B, and $C (ABC \neq 0)$. Fig. 7.1 illustrates the case $A = 1$, $B = 2/3$, and $C = 1/3$ (Dombre *et al.*, 1986). The plane $z = 0$ is repeatedly pierced by streamlines (such a plane is called the Poincaré plane). In the regions with ordered motions, subsequent points on this plane which are traces of the same streamline belong to a line that represents a cross-section of a certain two-dimensional surface. Such surfaces are called KAM (Kolmogorov-Arnold-Moser) tori. According to the KAM theory, the trajectories of the vector field always lie on tori or rings (near boundaries), provided its curl is not collinear with the field

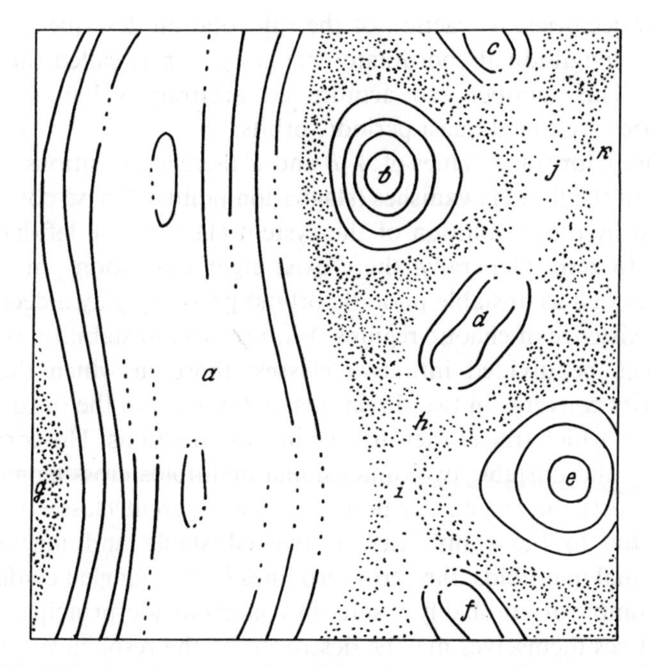

Fig. 7.1. Schematic view of a cross-section of the Poincaré plane filled by the trajectories of a stochastic dynamic system. The regions $a-f$ are filled by the cross-section of invariant tori. In other parts (g and h), traces of stochastic trajectories can be seen. The trajectories do not uniformly fill the invariant tori and the stochastic regions. This makes the character of trajectories in the regions $i-k$ uncertain.

itself, i.e. $\nabla \times \mathbf{v} \neq \lambda \mathbf{v}$. In the considered case this condition is not fulfilled and, apart from the tori, there appear regions with chaotic streamlines. (The points shown in Fig. 7.1 trace a single trajectory that repeatedly pierces the plane $z = 0$.)

When one of the constants vanishes, e.g., $C = 0$, then the system has a first integral of motion $\Phi[x(t), y(t), z(t)] = \Phi_0 =$ const. and thus the streamlines lie on a surface $\Phi(x, y, z) =$ const. Therefore, the chaotic regions are absent. The chaos can arise only in three-dimensional flows.

Numerical and analytical studies (Dombre *et al.*, 1986) show that, first, there exist regions where the flow is dominated by motion in some chosen direction. Every such region has the shape of a rope parallel to one of the axes. Taking into account the direction of the flow, we see that there are six such principal rope-vortices in which the fluid moves along helical trajectories. A fraction of the full rotation described by a helix when the longitudinal coordinate changes by 2π is called the *rotation number*. This number can acquire an arbitrary value. The helical trajectories are called quasi-periodic orbits.

Secondly, for some values of A, B and C there are positions where the velocity of the flow (1) vanishes (stagnation points). These positions can be found by direct solution of the system (1) with the left-hand sides equated to zero. Generally, there exist eight stagnation points. These points (as well as unstable periodic orbits) probably play a decisive role for the existence of chaotic regions. With respect to stability, stagnation points can be divided into two classes: those in which the nearby trajectories converge in two directions and diverge in the third one, and those from which trajectories diverge in two directions. The corresponding stable and unstable two-dimensional manifolds cross along the line that connects the stagnation points of the different classes (zero–zero lines). Due to the complexity of twisted stable and unstable two-dimensional manifolds, the zero–zero lines form a tangled configuration whose complexity probably results in chaos outside principal vortices. The vortices themselves may be destroyed by the resonances that occur for rational rotation numbers; such a destructive role is played, e.g., by the resonance 3:1 (Dombre *et al.*, 1986).

7.2. THE ROLE OF CHANCE IN TRANSITION TO TURBULENCE

A steady flow of incompressible liquid, neglecting the effects of the finite speed of sound, is determined by the shape of the region occupied by the flow (usually, what matters is the characteristic region size, l[cm]), by a characteristic velocity $v[\mathrm{cm\,s^{-1}}]$, and by kinematic viscosity of the liquid $\nu[\mathrm{cm^2 s^{-1}}]$. These quantities can be combined into a single dimensionless combination known as the *Reynolds number*

$$Re = \frac{lv}{\nu}. \tag{2}$$

Flows with equal Reynolds numbers in geometrically similar regions are also similar (the similarity law of Reynolds). For instance, the solution of the problem of the flow of mercury around a sphere can be obtained from the solution of the analogous problem of flow of water by a simple rescaling of coordinates and velocities; the sphere diameter and the water velocity should be chosen so as to provide the same Reynolds number.

In principle, a steady flow $\mathbf{v}_0(\mathbf{r})$, as solution to the Navier-Stokes equation, exists for any value of the Reynolds number. However, experiments and mathematical analysis of particular cases have shown that such flow is not stable for any value of *Re*. When, due to the loss of stability, a regular steady velocity field turns into a tangled unsteady one (or steady only in the mean), it is said that a transition from laminar motion to the turbulent one occurs.

Formally, stability analysis can be accomplished as follows (Landau and Lifshitz, 1959). First, a steady solution $\mathbf{v}_0(\mathbf{r}; Re)$ of the Navier-Stokes equation

$$(\mathbf{v}_0 \cdot \nabla)\mathbf{v}_0 = -\nabla\left(\frac{p}{\rho}\right) + Re^{-1}\Delta\mathbf{v}_0 ,$$

$$\nabla\mathbf{v}_0 = 0 .$$

is found (as a rule, in analytical form with rare exceptions of numerical analysis), where p and ρ are the fluid pressure and density, respectively. Then weak perturbations $\mathbf{v}_1(t, \mathbf{r}, Re)$ and $p_1(t, \mathbf{r}, Re)$ are introduced which are described by a *linearized* equation with coefficients dependent on $\mathbf{v}_0(\mathbf{r})$:

$$\frac{\partial\mathbf{v}_1}{\partial t} + (\mathbf{v}_0 \cdot \nabla)\mathbf{v}_1 + (\mathbf{v}_1 \cdot \nabla)\mathbf{v}_0 = -\nabla\frac{p_1}{\rho} + \nu\Delta\mathbf{v}_1 .$$

Since coefficients of this equation do not depend on time, the time dependence of the solution is given by

$$\mathbf{v}_1 = \mathbf{f}(\mathbf{r}) \exp[\gamma t + i(\omega_1 t + \varphi_1)t] .$$

The growth rate γ and frequency ω of the developing flow are functions of the Reynolds number. The value of *Re* for which γ crosses zero and becomes positive is called the critical Reynolds number, Re_{cr}.

In contrast to the growth rate and the frequency determined by the linearized Navier-Stokes equation, the phase φ_1 of the perturbation remains undetermined and depends on the random initial conditions. This situation is expressed by saying that the flow has acquired one more degree of freedom. We emphasize that the basic steady flow has no degrees of freedom. This is a freedom won by chance!

Nonlinear effects eventually suppress the growth of $\mathbf{v}_1$, i.e., γ vanishes. However, ω_1 can remain non-vanishing and what arises is a periodic flow, stable for a given value of *Re*. With further increasing Reynolds number, this flow becomes unstable too. This new perturbation gives rise to a second frequency ω_2, independent of ω_1, and to a second degree of freedom φ_2. Geometrical interpretation of this flow is a path which winds around a torus with the frequency ω_1 corresponding to the rotation along the large circle and ω_2 being the frequency of rotation along the small circle. When these frequencies are incommensurate, the trajectory is unclosed.

L. Landau in 1944 and E. Hopf in 1948 have independently proposed the hypothesis (see Landau and Lifshitz, 1959) that for further increase of the Reynolds number, new frequencies and new degrees of freedom (phases) will appear according to this very scheme. The difference between subsequent critical Reynolds numbers decreases along this sequence, and the scale of every new mode of motion is smaller than that of the preceding one, i.e., the flow becomes more and more tangled and turbulent. According to this model, turbulence is actually a superposition of a large number of modes with independent (and incommensurate) frequencies and random phases. We see that chance plays a decisive role in the Landau-Hopf scheme. Nonlinearity is invoked here only in order to saturate every next periodic perturbation. Actually, nonlinear effects can play a more important role and can even be fatal for this model. The point is that the interaction between a newborn mode with frequency, e.g., ω_2 and the preceeding periodic motion with frequency ω_1 can lead to synchronization of these modes with locking of the mode ω_2 by the mode ω_1, thus producing a one-frequency periodic flow with frequency ω_1. The theory of dynamic systems predicts that such synchronizing is possible for a wide range of relevant parameters. However, the Landau-Hopf scenario of the route to turbulence is not entirely rejected and in principle flows can be made turbulent through random excitation of new degrees of freedom.

Nonlinear effects can destroy quasi-periodic regimes, not only by simplifying them by synchronizing modes but also by adding complexity when in the phase space of the system a strange attractor appears, i.e., the attracting set on which the trajectories are very tangled and unstable. As shown by Ruelle and Takens (1971), a strange attractor cannot appear on a two-dimensional torus (unstable trajectories cannot belong to a closed two-dimensional surface!) and its appearance is possible only after the three-periodic motion (three-torus) is destroyed.

It would seem that the Ruelle-Takens scenario of turbulent transition is dominated by nonlinearity while chance plays a negligible role. Actually, the possibility of stochastization is based on the high sensitivity of the trajectory behaviour to the initial conditions. Clearly, in a stable flow, weak modification of the initial conditions leads to weak changes of trajectories after a finite period. In contrast, in unstable motion on a strange attractor, any initial inaccuracy in fixing the initial conditions grows exponentially. Therefore, when the initial conditions are exactly fixed, the behaviour of the system is absolutely predictable and free of chaos. The point is that uncontrollable weak initial uncertainty turns into unpredictability of the trajectories' behaviour after a finite time. Thus, an ultimate random behaviour is a consequence of strong enhancement of the initial uncertainties.

7.3. THE DEVELOPED TURBULENCE

For sufficiently large Reynolds numbers, practically every three-dimensional hydrodynamic flow becomes turbulent. Properties of the developed turbulence are independent of the particular way of turbulent transition and therefore can be understood from very general considerations. This approach was first proposed by L. Richardson and acquired clear quantitative presentation in the works of A.N. Kolmogorov and A.S. Obukhov (see Monin and Yaglom, 1971; Landau and Lifshitz, 1959).

According to this classical theory, the turbulence is driven, i.e., the energy is injected, at some basic scale l. The reader can imagine a liquid stirred by blades of diameter l comparable to the size of the volume occupied by the liquid. After the period l/v_l, where v_l is the velocity at the scale l, because of strong nonlinear interactions a basic vortex gives rise to vortices of half this scale. Say, for the vortex flow $v_x = v_l \sin y/l$, $v_y = v_l \sin x/l$, the nonlinear term $((\mathbf{v} \cdot \nabla)\mathbf{v})_x \sim v_y \, \partial v_x/\partial y$ in the Navier-Stokes equation gives rise to the motion of the scale reduced by a factor of two and of an amplitude comparable to v_l. The resulting pair of vortices is subject to the same nonlinear effects and they disintegrate into four vortices, etc. The large value of the Reynolds number $Re = lv/\nu \equiv (l^2/\nu)/(l/v)$ implies that the characteristic fragmentation time is shorter than the dissipation time l^2/ν at the considered scale. This allows one to suppose that the kinetic energy is not dissipated in the process of fragmentation of vortices but is transferred from motions of larger scales to those of smaller scales, i.e., the spectral energy flux is conserved:

$$\varepsilon = \frac{v_\lambda^2}{\tau} = \frac{v_\lambda^3}{\lambda} = \text{const.},$$

where λ is the motion scale. This relation leads to the well known *Kolmogorov spectrum*

$$v_\lambda \sim (\varepsilon\lambda)^{1/3} \, . \tag{3}$$

The vortex fragmentation proceeds further until the scale of smaller vortices becomes so small that λ/v_λ becomes of the order of the dissipation time λ^2/ν at this scale. By equating these two time scales and substituting $v_\lambda \propto \lambda^{1/3}$, one obtains the characteristic dissipation scale

$$\lambda_D \approx l Re^{-3/4} \, . \tag{4}$$

The interval of scales (l, λ_D) is called the inertial range. The arguments above give a qualitative form of the energy spectrum of the turbulence (Fig. 7.2), i.e., the amount of energy per unit mass, in terms of the inverse scale $K = 2\pi/\lambda$ at the given scale,

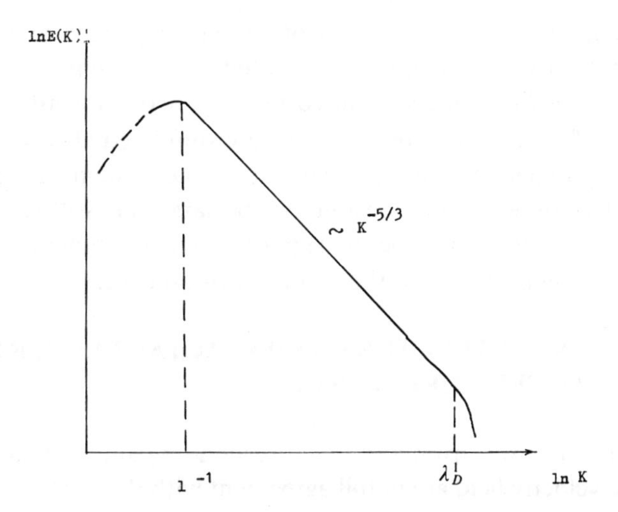

Fig. 7.2. The energy spectrum of the Kolmogorov turbulence.

$$\int^{K} E(K)\,dK \sim v_K^2 \sim K^{-2/3} .$$

The large-scale tail of the spectrum shown dashed in Fig. 7.2 depends on the kind of forcing and is not of the universal character.

In fact, the velocity of a turbulent flow is a random function of time and position. Therefore, a more consistent description should be in terms of statistical averages. Then the expressions given above characterize the autocorrelation function $\langle v_i(\mathbf{x} + \mathbf{r}) \cdot v_i(\mathbf{x}) \rangle \equiv \xi(r) v_0^2$ of the velocity field which is homogeneous and isotropic and has a vanishing mean velocity on average. Clearly, $\xi(0) = 1$. At scales below λ_D the velocity field is regular, i.e., the autocorrelation function behaves as $\xi = 1 - 1/2\,(r/\lambda_T)^2$. The characteristic scale λ_T (the Taylor microscale) that appears here is determined by the second derivative of the autocorrelation function, $\lambda_T = [-\xi''(0)]^{-1/2}$, and characterizes only the scale of autocorrelation. The inertial range, where $\xi - 1 \sim r^{2/3}$, ends at the scale λ_D which is much smaller than λ_T.

A straightforward interpretation of the given picture of turbulence implies that turbulence can be represented as an ensemble of approximately $(l/\lambda_D)^3 \sim Re^{9/4}$ independent vortices, or, in other words, degrees of freedom. In fact, this picture is excessively simplified. The vortices can be mutually dependent or correlated, and the system can possess a strange attractor whose dimension can be large but still smaller than $Re^{9/4}$. Determining the number of degrees of freedom remains one of the most interesting problems of the theory of turbulence.

7.4. INTERNAL STRUCTURE AND FRACTAL PROPERTIES OF DEVELOPED TURBULENCE

The Kolmogorov spectrum and its corresponding autocorrelation function of the velocity field are in full agreement with the experimental data on laboratory and atmospheric turbulence (see, e.g., Monin and Yaglom, 1971).

However, a naive extrapolation of the Kolmogorov scaling law to higher statistical moments,

$$\langle v_\lambda v_\lambda \ldots v_\lambda \rangle \sim \lambda^{p/3}, \quad \lambda_p < \lambda < l$$

proves to be incorrect! The discussion of the corresponding experimental results and the theoretical considerations are presented by Frisch (1985). According to the experiments, what is observed is the index ζ_p shown in Fig. 7.3 obtained for $Re = 852$ by Anselment *et al.* (1970), rather than the index $p/3$ as expected from the Kolmogorov law. This result can be presented in a more formal way as follows (Frisch and Parisi, 1985).

The Kolmogorov law $v_\lambda \sim \lambda^{1/3}$ implies that the velocity field has 1/3 derivative, in the sense of Hölder, in the limit $\lambda \to 0$, i.e., for $Re \to 0$. In reality, the number of the Hölder derivatives is not a spatially uniform value. Denoted by $S(\sigma)$ the set of positions where the velocity field has less than σ derivatives. Let $d(\sigma)$ be the Hausdorff dimension of the set S (see Chapter 5). The probability that for $|\mathbf{x} - \mathbf{y}| \to 0$ the difference $|\mathbf{v}(\mathbf{x}) - \mathbf{v}(\mathbf{y})|$ is of the order of $|\mathbf{x} - \mathbf{y}|^\sigma$ is equal to the ratio of the volume of the set $S(\sigma)$ to the volume of the whole space, i.e., to $|\mathbf{x} - \mathbf{y}|^{3-d}$. Hence,

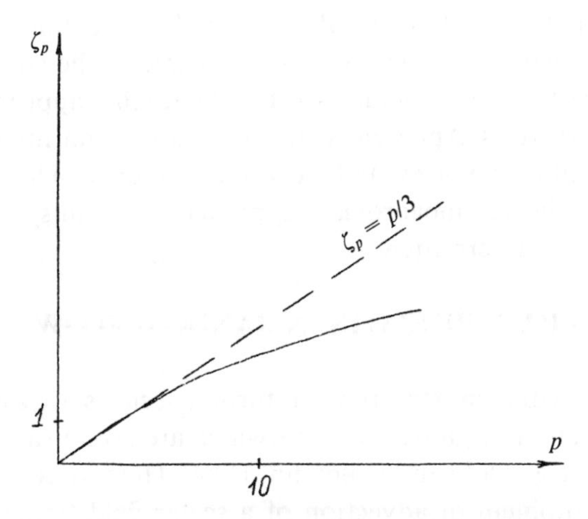

Fig. 7.3. The behaviour of higher moments of turbulence (dashed line) deviates from the Gaussian-Kolmogorovian picture (solid line).

$$\langle v_\lambda^p \rangle \sim \int \lambda^{p\sigma+3-d(\sigma)} d\sigma .$$

Estimating this integral by the saddle-point method, one obtains

$$\zeta_p = \min_\sigma [p\sigma + 3 - d(\sigma)] . \tag{5}$$

This estimate allows one to find the dimension of the point set from the observed values of ζ_p where the velocity field has less than σ derivatives, $d = 3 - \min_p (\zeta_p - p\sigma)$ (see Fig. 5.6). This dependence of d on σ should be interpreted in the following way. The majority of the positions in the order of the Hölder derivative is close to 1/3. There are some positions at which the order of the Hölder derivative is lower, down to 1/10, and some positions where the velocity field is almost smooth with the Hölder derivative being of the order 0.9. Thus, the simplified (Kolmogorov) picture levels off and hides a differential richness of the turbulent velocity field in the limit of the infinitely large Reynolds number.

In physical terms, these results imply that the homogeneous and isotropic turbulence is structured. Some degree of the structuredness is suggested by the very picture of the hierarchy appearing and the fragmenting vortices. Apparently, the autocorrelation function behaves differently within a vortex and near its boundary. This difference is neglected by the Kolmogorovian approach. And this difference is a manifestation of intermittency.

7.5. TEMPERATURE SPOTS IN RANDOM FLOW

Studies of the coherent structures in turbulent flows are associated with complicated and intriguing problems which are now a subject of intense research (see, e.g., Dold and Eckmann, 1984). Here we consider a similar but simpler problem of advection of a scalar field (temperature) by a given random flow. The simplicity of this problem is due to its linearity but, with respect to the effects of randomness, it seems to be no less rich than the problem of coherent structures. In a turbulent flow of liquid with heat sources, the temperature is a random function of time and position. Usually the subject of analysis is the behaviour of a locally averaged temperature. In this case the effects of random flow are reduced to advection at the mean velocity and turbulent temperature diffusion. However, the mean temperature does not provide a complete description of all the effects. Even for a moderate mean temperature, temperature fluctuations sometimes can be so high that they lead to critical phenomena, e.g., local boiling.

In this discussion, we reveal and describe the causes of deviations of the temperature from its mean value in a flow around the heat sources. These temperature inhomogeneities have the form of spots whose scale and lifetime are determined by the velocity field and by the Peclet number $Pe = lv/\kappa$, where κ is the molecular heat conductivity. As an application of the developed theory, we consider boiling in an underheated liquid in a nuclear reactor (Dvorina *et al.*, 1989).

Let us begin with the analysis of temperature fluctuations. Consider the problem of heat conduction in a random incompressible flow whose velocity is $\mathbf{v}(t, \mathbf{r})$ in the presence of heat sources that release the energy $\rho c_p Q(t, \mathbf{r})$ per unit volume per unit time:

$$\frac{\partial T}{\partial t} + (\mathbf{v} \cdot \nabla) T = \kappa \Delta T + Q\,, \quad T(0, \mathbf{r}) = T_0(\mathbf{r})\,. \tag{6}$$

The velocity field and heat sources are characterized by the average values u and Q, respectively, and by random components with r.m.s. deviations σ_v and σ_Q, respectively. The temperature is also a random field with the mean value $\bar{T}(t, \mathbf{r})$ and fluctuations $T' = T - \bar{T}$.

Let us consider an interesting and complicated case of large Peclet numbers using the Lagrangian formulation of the problem. The essence of this formulation is the following.

Put $\kappa = 0$. Then the evolution of the temperature field reduces to advection of the initial distribution by a fluid particle with additional heating by the sources encountered along the trajectory. Thus,

$$T(t, \mathbf{r}) = T_0(\xi_t) + \int_0^t Q(t - s, \xi_s)\, ds\,, \tag{7}$$

where ξ_t is the initial position of a particle that passes through the considered point r at given moment t. For a given velocity field $\mathbf{v} = d\xi_t/dt$, the Lagrangian coordinate can be obtained from the equation

$$\xi_t = \mathbf{r} - \int_0^t \mathbf{v}(t - s, \xi_s)\, ds\,. \tag{8}$$

When molecular heat conduction is taken into account, the evolution of the temperature field cannot be described as advection along the trajectories of fluid particles. However, one can introduce the notion of a fictitious random path which differs from (8) by adding the term $(2\kappa)^{1/2}\mathbf{W}_t$ to the Wiener (Brownian) random process responsible for the random wandering associated with the diffusion process (see Chapter 3). Then a physical solution is obtained after averaging of (7) over a bundle of random paths that reach the position r at the moment t. Based on this solution, one can obtain and solve deterministic equations for the mean temperature and the correlation function of temperature fluctuations (Zeldovich *et al.*, 1988).

As can be seen from the solution (7), there are three sources of temperature fluctuations in the flow: fluctuations in the initial distribution $T_0(\mathbf{r})$, fluctuations of heat supply, and the random character of the motion of fluid particles advecting heat. The last case is understood clearly when the reader imagines that a given position is reached by a relatively "hot" particle that has visited a larger number of heat sources.

These temperature fluctuations are spatially organized in spots whose characteristic size and shape are determined by the initial conditions, the distribution of the heat sources and the correlation properties of the velocity field.

Let us estimate the lifetime of the temperature spots. At first sight, under the action of the turbulent heat conduction, any spot would quickly spread out in time l^2/κ_T, where l is the basic scale of turbulence and κ_T is the turbulent heat conductivity. However, apart from being proportional to l^2, the spreading time also depends on the spot size through $D(l)$. For a small spot, the spreading is for a long period determined by (small) molecular heat conductivity, and only when the spot size becomes comparable to the basic turbulent scale l is the (strong) turbulent conduction switched on.

An estimate of the characteristic lifetime of a spot follows from the expression for evolution of the mean squared distance between two fluid particles that advect heat when moving along the random paths:

$$\frac{d\bar{r}^2}{dt} = 2D(\bar{r})\,. \tag{9}$$

For a short-correlated flow, it can be shown that the diffusion coefficient is approximated by $D = \kappa + \kappa_T(r/l)^2$ for $r \leqslant l$ and $D = \kappa_T$ for $r \geqslant l$ (Molchanov *et al.*, 1985). Integration of (9) from some small initial distance r_0 to l yields

$$t_s \simeq \frac{l^2}{\kappa_T} \ln \frac{\kappa_T}{\kappa + (r_0/l)\,\kappa_T} = \frac{l^2}{\kappa_T} \ln \frac{Pe}{1 + (r_0/l)\,Pe}\,. \tag{10}$$

Derivation of this expression implicity assumes that the velocity field has a statistical distribution close to the Gaussian one. When this is not true and there exist coherent structures in the flow, the lifetime of temperature spots can exceed the value given by (10). The reason for this is the fact that within such structures, the turbulent transfer effects are replaced by molecular heat conduction and viscosity. Therefore, for large Peclet and Reynolds numbers these structures would have very long lifetimes. The role of such coherent flow structures in heat conduction near various boundary surfaces has been discussed by Kirillov (1986). We only note that a coherent structure does not necessarily imply excess fluid temperature. Thus, the probability of formation of temperature spots associated with the stagnation-type flow structures is given by the product of the probability of development of the structure and the probability of development of temperature fluctuations in a given structure. The appearance of coherent structures in a turbulent fluid is associated either with transitional regimes from a laminar flow to the turbulent one (Section 7.2) for moderate Reynolds numbers, or with small scale effects for large Reynolds numbers (Section 7.4).

Note that the time t_s given by (10) also determines the time of diffusional spreading of a bundle of random paths, i.e., the limit of applicability of the Lagrangian solution (7).

Now we can estimate the amplitude of temperature fluctuations. Deviations of the temperature from the mean value, $T' = T - \bar{T}$, caused by fluctuations in the frequency of approach of fluid particles to heat sources and by fluctuations of heat supply are described by the second term in the right-hand side of (7). When a path visits a large number (N, on average) of the sources, the deviation, being a sum of a large number of random independent contributions, is given by

$$T' = \bar{Q}\tau_H N^{1/2}\xi_1 + \sigma_Q \tau_H N^{1/2}\xi_2\,, \tag{11}$$

where ξ_1 and ξ_2 are random, independent Gaussian quantities with zero mean value and unit dispersion. Here τ_H is the residence time in the vicinity of a heat source which is considered constant for simplicity. Notice that fluctuations of τ_H can produce a significant effect, especially in the case where there are flow structures near a wall. During the period when the fluctuation T' is produced, the mean temperature also grows (due to $\bar{Q}$) having the total increment $\Delta\bar{T} = \bar{Q}\tau_H N$. Since the random quantities in (11) are independent, we can replace ξ_1 and ξ_2 by a single random quantity ξ, having added the dispersions. This brings us to the following expression for temperature fluctuations in the flow:

$$T' = \Delta\bar{T}\left(1 + \frac{\sigma_Q^2}{\bar{Q}^2}\right)^{1/2} \frac{\xi}{N^{1/2}}. \qquad (12)$$

The independent contribution of initial temperature inhomogeneities should be added to this expression.

We emphasize that we have not assumed that the fluctuations of the heat supply are Gaussian. The Gaussian distribution of T' arises because this is a sum of a large number of random contributions, i.e., because a fluid particle visits the places with various values of the random quantity Q several times. When the distribution of $Q - \bar{Q}$ strongly differs from the Gaussian one and N is not very large, the expression (11) ceases to be applicable at least because the Gaussian distribution cannot develop. In this case T is determined by the maximal value of Q encountered along a trajectory. In other words, a strong deviation of heat supply leads to the formation of a pronounced temperature spot whose lifetime is t_s.

The average number of approaches of a fluid particle to the heat sources can be estimated as the ratio of the circulation time τ_p of travel of the fluid along the system of heat sources to the time of random walk between the neighbouring sources:

$$N = \frac{\tau_p}{l^2/\kappa_T}. \qquad (13)$$

It is presumed here that the lifetime (10) of temperature spots exceeds the circulation time. Substitution of (13) into (12) gives the r.m.s. temperature deviation:

$$\sigma_T = \Delta\bar{T}\left(1 + \frac{\sigma_Q^2}{\bar{Q}^2}\right)^{1/2} \left(\frac{l}{\kappa_T \tau_p}\right)^{1/2}. \qquad (14)$$

The turbulent temperature conductivity can be estimated, e.g., according to the mixing length theory as $\kappa_T = \varepsilon l \sigma_v$, where $\varepsilon = 0.3 + 0.01$ is the numerical coefficient. The circulation time is, evidently, $\tau_p = L/u$.

In the opposite case, for $t_s < \tau_p$ we have $N \sim \tau_s (\kappa_T/l^2)$ and in (12) one should not consider the total increment of the temperature but rather the increment produced in the period t_s:$(\Delta \bar{T})_s = \Delta \bar{T} t_s/t$. Then

$$\sigma_T = \Delta \bar{T} \left(1 + \frac{\sigma_Q^2}{\bar{Q}^2}\right)^{1/2} \left(\ln \frac{Pe}{1 + (r_0/l)\, Pe}\right). \tag{15}$$

Many subsequent circulations of the fluid through the considered volume V can be considered as independent trials. Therefore, the number of realizations of the random quantity ξ in (14) depends not only on the number of independent turbulent cells within the volume V, but also on the number of fluid circulations which leads to the increase of the effective volume. The random quantity ξ can acquire generally an arbitrarily large value. But it is clear that only those events are realizable for which the product of their probability $P(\xi > \xi_0)$ and the total number of independent trials is of order unity, i.e., $P(t/\tau_p)(V/l^3) \sim 1$, where t is the duration of the experiment and τ_p is the time of a single circulation. Substituting the Gaussian distribution of ξ leads to the relation

$$\xi_0 = \sqrt{2 \ln \left(\frac{tV}{\tau_p l^3}\right)}$$

for the standard deviation. This allows one to estimate the maximal amplitude of the temperature fluctuation during the total duration of the experiment:

$$|T'|_{\max} \simeq \sigma_T \xi_0 \,. \tag{16}$$

As an example, consider the phenomenon of underheated boiling; this phenomenon is important for the problem of the safety of nuclear reactors (Tong, 1981). This problem is, in turn, associated with the estimation of the required amount of material for the envelops of the heating elements.

The underlined boiling is the liquid boiling at the surface of the heating elements when the liquid temperature outside a thin layer at the surface is below the boiling point. In contrast to the one-phase convective flows, the underheated boiling leads to the formation of vapour bubbles and, thus, stronger temperature perturbations and to the activation of exchange processes, e.g., the deposit of the products of corrosion. The underheated boiling produces a thermally non-equilibrium two-phase flow which is far from thermal equilibrium with coexisting vapour bubbles and the liquid whose temperature is below the boiling point.

A random character of boiling can be naturally associated with temperature fluctuations in a flow of the heat-transfer agent. For order-of-magnitude estimates of the parameters of temperature spots we adopt the following characteristic parameters typical of the problem of water cooling of a system of heating rods: $Pe = 7 \times 10^4$, $l = 10^{-2}$ m, $L = 2.5$ m, $u = 1$ ms^{-1}, $\sigma_v/u = 0.1, \varepsilon = 0.3$. Then the ratio of the lifetime of a temperature spot to the crossing time is given by

$$\frac{t_s}{\tau_p} = \frac{1}{3}\frac{l}{L}\frac{u}{\sigma_v} \ln Pe \simeq 1.5 \, .$$

Thus, all the spots survive during the time of the passing of water through the system of heating elements.

Temperature fluctuations associated with randomness of the flow can be obtained by putting $\sigma_Q = 0$ in (14) and by adopting a smooth initial temperature distribution. They turn out to be considerable:

$$\frac{\sigma_T}{\Delta \bar{T}} \simeq \left(\frac{lu}{\varepsilon L \sigma_v}\right)^{1/2} = 0.35 \, . \tag{17}$$

This ratio decreases with decreasing "equivalent diameter" of the system l and decreasing of the ratio u/σ_v. For instance, for $l = 3 \times 10^{-3}$ m we obtain $\sigma_T/\Delta\bar{T} = 0.2$. Notice that initial temperature inhomogeneities and heat supply fluctuations can produce comparable temperature fluctuations (Tong, 1981).

Let us estimate the number μ of bubbles in a given fluid volume V. Since the temperature fluctuations obey the Gaussian statistical law, it is sufficient to estimate the number of crossings of the level $T_s - \bar{T}$ by the Gaussian field T', where T_s is the boiling point. Denoting $\xi_s = (T_s - \bar{T})/\sigma_T$ we have for large ξ_s:

$$\mu = \frac{V}{(2\pi l)^3}\,\xi_s^2 \exp\left(-\frac{\xi_s^2}{2}\right) \tag{18}$$

(Nosko, 1969). This function of ξ_s is maximum for $\xi_s = 2^{1/2}$. For smaller ξ_s one can approximate μ by the number of turbulent cells within the volume V, i.e., the whole fluid is boiling. If expression (17) is applicable, this implies that total boiling occurs when the average liquid temperature is 1.5 times smaller than the boiling temperature.

CHAPTER

EIGHT

GENERAL CONCEPTION OF INTERMITTENCY

We have already mentioned intermittency in the discussions above. Intermittency plays an essential role in the considered aspects of applications of the probability theory to physics. For instance, in intermittent distributions of matter, its mass is mainly concentrated within widely spaced concentrations in which the density exceeds the average one. Such kind of matter distribution is characteristic of the universe where the density within stars and planets is incomparably greater than the average density of galaxies while the latter is much greater than the average density of the universe.

This chapter is devoted to some elementary questions and this discussion helps us to reveal the nature of intermittency and introduce relevant concepts. We begin with the discussion of a random quantity and proceed to that of intermittency in a random field. The chapter concludes with the discussion of the role of nonlinearity.

8.1. INTERMITTENCY OF A RANDOM QUANTITY

In practically every physical experiment a random quantity ξ appears. Usually such random quantities are believed to be distributed according to the normal, or Gaussian, law and thus completely determined by the mean value $\langle\xi\rangle$ and the dispersion σ^2. Majority of the values of the random quantity are close to the mean value and reside within the region whose radius is of the order of the r.m.s. deviation equal to σ, i.e., to the square-root of the dispersion. The quantity ξ can acquire values that strongly differ from $\langle\xi\rangle$ but such strong deviations have very low probability indeed. This property is the basis for the success of the least squares method, which is widely used in processing experimental data. The Gaussian random quantities usually arise as a sum of a large number of small, independent or weakly inter-dependent random variables which are approximately identically distributed. The Gaussian character of such sums is guaranteed by the central limit theorem.

However, consider a random quantity of the multiplication type which is a product of a large number of identically distributed independent random quantities. Let ξ_j, $j = 1, \ldots, N$, acquire the values 0 and 2 with equal probabilities 1/2. Then the random quantity given by the product

$$\xi = \prod_{j=1}^{N} \xi_j = \xi_1 \xi_2 \ldots \xi_j \ldots \xi_N$$

vanishes for almost all realizations of ξ_j. The only exception is the one realization where all ξ_j are equal to 2. The probability for this unique realization is very low: for large N, it is given by 2^{-N}. On the other hand, the value of ξ for this realization is very high and is equal to 2^N.

This random quantity ξ has an amazing statistical distribution. This is very different from the Gaussian one. All possible values of ξ are zeros except a unique one which is very large. However, this very value determines the mean value of the random quantity:

$$\langle \xi \rangle = \frac{\text{sum of all realizatons}}{\text{number of realizations}} = \frac{0 + 0 + 0 + \ldots + 0 + 2^N}{2^N} = 1\,.$$

The mean square

$$\langle \xi^2 \rangle = \frac{\text{sum of squared realizatons}}{\text{number of realizations}} = \frac{0 + 0 + \ldots + 0 + 2^{2N}}{2^N} = 2^N$$

exponentially grows with increasing N. The higher moments $\langle \xi^3 \rangle, \langle \xi^4 \rangle, \ldots, \langle \xi^p \rangle = 2^{(p-1)N}$ grow still stronger. The rate of growth of the statistical moments with N is given by

$$\gamma_p \equiv \frac{\log_2 \langle \xi^p \rangle}{N} = p - 1\,. \tag{1}$$

We see that the growth rate also grows with the order of the moment, p. In the limit $p \to \infty$ we have $\gamma_p \to p$.

Random quantities of the considered type are called *intermittent random quantities.* We emphasize that this notion is associated with large values of N, i.e., with large numbers of multipliers. Similar to a Gaussian random quantity which is typical of the sum of many random numbers, the intermittent random quantity seems to be typically a result of the multiplication of many random numbers.

The simple example considered above may seem to degenerate due to the dominance of zeros in ξ_j. However, the presence of intermittency is independent of these zeros. Consider another random quantity ζ_j distributed around unity. Then the logarithm of the product of realizations,

$$\ln \zeta = \ln \zeta_1 + \ln \zeta_2 + \ldots + \ln \zeta_n$$

is a sum of many random numbers distributed around zero. Therefore, for large N we have $\ln \zeta \sim N^{1/2} \eta$, where η is a Gaussian random quantity with zero mean value. Thus,

$$\zeta \sim \exp(N^{1/2} \eta)$$

which is a log-normal distribution. The quantity η has values mainly from the interval $(-\mu\sigma, \mu\sigma)$, where σ is the standard deviation of η and $\mu \sim 1$ is a numerical coefficient. For $\eta < 0$, the value of ζ is exponentially close to zero, while for $\eta > 0$ values of ζ are large, of the order of $\exp(N^{1/2}\sigma)$. The mean value of a log-normal quantity is

$$\langle \zeta \rangle = \int \zeta(\eta) p(\eta) d\eta = \int \exp(N_\eta^{1/2}) \exp\left(-\frac{\eta^2}{2\sigma^2}\right) d\eta \sim \exp\frac{N\sigma^2}{2}.$$

In contrast to an individual realization, the mean value grows monotonously as an exponential function of N. Higher statistical moments also grow exponentially:

$$\langle \zeta^p \rangle \sim \exp\left(p^2 \frac{N\sigma^2}{2}\right).$$

The growth rate of the p-th moment is given by

$$\gamma_p = \lim_{N \to \infty} \frac{\ln\langle \zeta^p \rangle}{N} = p^2 \frac{\sigma^2}{2} \qquad (2)$$

and grows infinitely with p.

Even though here the contrast between an individual realization and the averaged characteristics is not so pronounced as in the former example, this more realistic example preserves all the typical features of intermittency.

Strictly speaking, evaluation of the statistical moments in the latter example based on the log-normal distribution, which is only an asymptotic distribution, is not completely consistent (Novikov, 1971; Mandelbrot, 1972). It is more correct to evaluate first $\langle \zeta^p \rangle$ for any given N and only then take the limit $N \to \infty$. However, the values of η of the order of $p^2 \sigma^2 N^{1/2}$ make a dominant contribution to the integral $\langle \zeta^p \rangle$, while a true distribution of η is close to the Gaussian one for $|\eta| < cN^{1/2}$, where c is a constant to be determined by the particular true distribution of η. Therefore, the expression (2) is valid when $p^2 \sigma^2 < c$, i.e., when $p^2 \sigma^2$ is sufficiently small. When the individual multipliers ζ_i do not acquire large values with considerable probability, the growth rate must be close to (2) even for large values of $p^2 \sigma^2$. Notice that this difficulty associated with an enhanced influence of rare strong deviations does not arise in the evaluation of moments of an additive random quantity. This is because the log-normal distribution for a multiplicative random quantity is not as universal as the normal Gaussian distribution of an additive one. In this sense, analysis of a multiplicative random quantity cannot be reduced to simply taking the logarithm.

8.2. EVOLUTION OF A RANDOM QUANTITY

Additive random quantities naturally arise in the processing of experimental data. Meanwhile, multiplicative random quantities arise in an equally natural way in the solution of evolutionary problems, where N plays the role of time. Of course, some evolutionary problems generate additive random quantities as well. Consider a simple example of evolution of ψ governed by the equation

$$\frac{d\psi}{dt} = \xi(t),$$

where $\xi(t)$ is the random quantity with, say, Gaussian distribution with dispersion σ^2 and with rapid decay of time correlation. For the sake of simplicity consider ξ that *renovates* after every period of duration τ, i.e., over the time intervals $[0, \tau)$, $[\tau, 2\tau)$, . . . its values are independent and identically distributed. Then the solution is given by a time integral which reduces to a sum of independent random terms. For $t \gg \tau$ the central limit theorem applies to this sum, i.e.,[a]

$$\psi(t) \sim \langle\xi\rangle t + \tau\sigma\left(\frac{t}{\tau}\right)^{1/2}\eta,$$

where η is the Gaussian quantity with zero mean value and unit dispersion.

When $\langle\xi\rangle \neq 0$, the true solution approaches the mean value with the growth of t, i.e., *self-averaging* occurs. For $\langle\xi\rangle = 0$ we see that $\langle\psi\rangle = 0$ and the solution normalized by $t^{1/2}$ *converges to a limit distribution*, Gaussian in this case. The relative strength of fluctuations does not grow with time and even decreases for $\langle\xi\rangle \neq 0$. Formally, this property can be expressed as splitting of higher moments,

$$\langle|\psi|^{p+q}\rangle \sim \langle|\psi|^{p}\rangle\langle|\psi|^{q}\rangle$$

i.e., higher moments reduce to products of the lower ones. In other words, for large t, knowledge of the mean value and mean square is sufficient for a full description of ψ.

[a] This result is valid even for much less restrictive properties of ξ (Monin and Yaglom, 1967; Taylor, 1921).

However, these two averages are obviously insufficient for the description of solutions to an evolutionary equation of unstable type,

$$\frac{d\psi}{dt} = \xi(t)\psi\,, \tag{3}$$

whose solutions have the multiplicative nature:

$$\psi(t) = \psi_0 \exp \int_0^t \xi(s)\,ds \underset{t \gg \tau}{\approx} \prod_{n=0}^{t/\tau} \exp \int_{n\tau}^{(n+1)\tau} \xi(s)\,ds\,,$$

i.e.,

$$\ln\psi \underset{t \gg \tau}{\approx} \langle\xi\rangle t + \tau\sigma\left(\frac{t}{\tau}\right)^{1/2}\eta\,, \tag{4}$$

This solution is, thus, an intermittent random quantity. The fluctuations of $\psi(t)$ grow with time as $\exp[\tau\sigma(t/\tau)^{1/2}\eta]$. More exactly, this exponential function describes the growth of the ratio of two individual realizations of ψ which correspond to two different realizations of η. Now the mean value of ξ and its mean square characterize only the logarithm of ψ at large times.

We stress again that, strictly speaking, expression (4) is suitable for evaluating the mean values $\langle\psi^p\rangle$ only for small values of $\tau\sigma$. When the product $\tau\sigma$ tends to zero, the second term on the right-hand side of (4) does not tend to zero or to infinity only when the standard deviation σ of ξ is related to the renovation time τ by $\sigma \sim \tau^{-1/2}$. This case is known as short time-correlations approximation. In this case, $\xi(t) - \langle\xi\rangle$ is proportional to the Brownian (Wiener) random process W_t for which $\langle W_t\rangle = 0$ and $\langle W_t^2\rangle = t$. We often use this approximation below and in Section 8.4 we define in a more precise way the limiting transition $\tau \to 0$.

8.3. A RANDOM MEDIUM

The phenomenon of intermittency, which manifests itself as the occurrence of rare but intense peaks in the behaviour of a random quantity, looks to be highly degenerated and rare and thus of no importance for, e.g., an ordinary stockbroker. However, some hardly probable events lead to such catastrophic consequences that they dramatically change even everyday life.

Random media (random fields) that consist of a continuum of random quantities offer much wider possibilities for the realization of rare events.

Consider for example a medium with a force field described by a random potential $U(x, \omega)$ (Zeldovich *et al.*, 1985). The parameter ω enumerates the realizations of the potential so that for a fixed ω the potential is an ordinary deterministic function of position. Specifically, consider U with a Gaussian distribution with zero mean value and the dispersion σ^2. The potential is a random field of an additive type and it can be represented as a series of unphased Fourier modes.

In such a potential the matter has the equilibrium concentration given by the Boltzmann law[b]

$$n = n_0 \exp\left(-\frac{U}{kT}\right). \tag{5}$$

This distribution is not Gaussian because the dependence on U is nonlinear. This is obvious for $\sigma/kT \gtrsim 1$ but it would seem natural that for $\sigma/kT \ll 1$ the distribution is linear:

$$n = n_0\left(1 - \frac{U}{kT}\right), \quad \langle n \rangle \approx n_0$$

and the Gaussian statistics prevail.

Indeed, let us find the value of the potential corresponding to the most probable concentration of the matter. Since the probability density of n is given by

$$p(n(U)) = n_0 \exp\left(-\frac{U}{kT}\right)p(U) = n_0 \exp\left(-\frac{U}{kT} - \frac{U^2}{2\sigma^2}\right),$$

[b] Such a law arises not only in thermodynamics but also in chemical kinetics where the activation energy plays the role of U.

its maximum value $p_{\max} = n_0 \exp(\sigma^2/2k^2T^2)$ is reached for $U_{\max}/\sigma = -\sigma/kT$. However, a detailed analysis of higher statistical moments

$$\langle n \rangle = n_0 \exp\frac{\sigma^2}{2k^2T^2},$$

$$\langle n^2 \rangle^{1/2} = n_0 \exp\frac{\sigma^2}{k^2T^2},$$

$$\vdots \qquad \vdots$$

$$\langle n^p \rangle^{1/p} = n_0 \exp\frac{p\sigma^2}{2k^2T^2}$$

shows that the larger their order is, the larger they are: $\langle n^2 \rangle \gg \langle n \rangle^2$, $\langle n^4 \rangle \gg \langle n^2 \rangle^2$, etc. In other words, successive mean values are determined not by the most probable value of n, σ/kT, but rather by $p^{1/2}\sigma/kT$. Hence, the linear approximation does not provide a field description even for small values of σ/kT. This progressive growth of statistical moments with their order can be explained only by a much more pronounced dominance of rare intense peaks in the concentration distribution. In principle, high peaks are also present in the Gaussian potential where $\langle U^p \rangle^{1/p}$ grows as $p^{1/2}$. But this kind of growth is by far weaker than the corresponding exponential growth of the moments of the concentration. What is important is the fact that weak traces of intermittency in the Gaussian potential, negligible by themselves, are greatly enhanced and become dominant in the distribution of the concentration (5) that depends nonlinearly on U.

On the other hand, the concentration is a solution to the following linear equation with a random coefficient $\mathbf{v}$ (velocity field)

$$\nabla(\kappa \nabla n - n\mathbf{v}) = 0\,,$$

which is the stationary diffusion equation for a moving medium and where in the mobility approximation we have $\mathbf{v} = (\kappa/kT)\nabla U$, with κ as the diffusion coefficient. This fact makes the basic difference between the intermittent structures and the structures which are the subject of synergetics and which arise due to processes described by nonlinear equations.

Linear scalar equations with random breeding and diffusion are typical of a number of problems from biology and kinetics of chemical and nuclear reactions (Mikhailov and Uporov, 1984). Zeldovich (1983) also considers some relevant nonlinear problems. Let us consider the phenomenon of intermittency using a simple equation

$$\frac{\partial \psi}{\partial t} = \kappa \Delta \psi + U(x, \omega)\psi, \qquad \psi(x, 0) = \psi_0(x), \tag{6}$$

as an example. Here the potential U has the Gaussian distribution with a certain characteristic scale l of decay of spatial correlations. A solution to equation (6) can be written (Kac, 1973; Gelfand and Yaglom, 1956) in a form similar to the form of the solution of the simpler diffusionless equation (3):

$$\psi(\mathbf{x}, t) = M_{\mathbf{x}} \left[\exp\left(\int_0^t U(\xi_s) ds \right) \psi_0(\xi_s) \right], \tag{7}$$

where the symbol $M_{\mathbf{x}}$ denotes averaging over all trajectories of the Brownian motion $\xi_s(\mathbf{x}) = (2\kappa)^{1/2} \mathbf{W}_s(\mathbf{x})$ that begin at the position ξ_t at the moment $s = 0$ and reach the position $\mathbf{x}$ at the moment $s = t$.

For any bounded non-negative initial function $\psi_0(\mathbf{x})$ and for any $k > 0$, the solution (7) with probability one asymptotically, for $t \to \infty$, grows as $\exp[6t\sigma^2 \ln(kt/l^2)^{1/2}]$. To be more exact, there exists a finite limit

$$\lim_{t \to \infty} \frac{\ln \psi}{t(\ln kt/l^2)^{1/2}} = (6\sigma^2)^{1/2}. \tag{8}$$

This kind of time dependence is explained by the fact that the dominant contribution to solution (7) is made by that trajectory which quickly encounters a high maximum of the potential. Let us estimate the height of this maximum. The region of the radius R (with $R \gg l$) contains $\sim (R/l)^3$ correlation cells. The probability that the potential reaches a given value U_0 in the cells is given by $P \sim \exp(-U_0^2/2\sigma^2)$. The condition $P(R/l)^3 \sim 1$ leads to the estimate

$$\max U \sim \left(6\sigma^2 \ln \frac{R}{l}\right)^{1/2} .$$

For a typical trajectory we have $R/l \sim (\kappa t/l^2)^{1/2}$ and we obtain a super-exponential growth $\psi \sim \exp[t(3\sigma^2 \ln(\kappa t/l^2)^{1/2})]$ even for a typical trajectory. In fact, still greater contribution does not come from a typical trajectory but rather from a less probable one called (by I.M. Lifshitz) the optimal trajectory. The shift from the initial position along the optimal trajectory is much greater, of the order of $R/l \sim \kappa t$ in a period t; due to this, the small statistical weight of this trajectory is compensated by a large factor $\max U$ (Zeldovich *et al.*, 1985).

We note that although the limiting value (8) is independent of molecular diffusivity, for $\kappa = 0$ the solution can only grow exponentially, as it follows from (6). This peculiarity is due to the dependence on κ of the time l^2/κ of the establishment of the super-exponential regime. Thus, the result depends on the order of two subsequent limiting transitions, either $t \to \infty$ and then $\kappa \to 0$, or $\kappa \to 0$ and then $t \to \infty$. Note also that the result (8) should not be interpreted as suggesting that diffusion κ enhances the growth of ψ. A more detailed analysis (Zeldovich *et al.*, 1987) shows that, taking into account further terms of expansion in diffusivity, the growth rate of ψ becomes a decreasing function of κ.

Even for $\kappa = 0$, the higher statistical moments of the field ψ behave super-exponentially. Indeed,

$$\langle \psi^p(\mathbf{x}, t) \rangle = \langle \exp(pUt) \rangle \psi_0^p = \psi_0^p \exp\left(\frac{p^2\sigma^2 t^2}{2}\right) .$$

The same law describes the evolution for $t \to \infty$ and $\kappa \neq 0$ as well (see Zeldovich *et al.*, 1985). Thus, the moments $\langle \psi^p \rangle^{1/p}$ grow much faster than the field ψ itself and the growth rate grows with the moment order. This progressive growth of statistical moments is associated again with the presence of sharp peaks in the solution $\psi(x, t)$, i.e., with the intermittency of the distribution of the field ψ. These peaks are present in any realization of the solution that corresponds to a given realization of the potential. In their positions they coincide with rare and high maxima of the potential U. For comparison, note that in the evolution of a random quantity, similar peaks appear only in some rare realizations.

Above we have considered an infinite medium. For a finite volume, the result reads $\psi \sim \exp(\max U, t)$, where $\max U$ is the random quantity, e.g., of the Gaussian type. For a Gaussian potential, the moments still grow as $\exp(p^2\sigma^2 t^2/2)$. However, if the potential is bounded from above[c] by some value $\bar{U}$, the result is $\psi_{\max} \sim \langle \psi^p \rangle^{1/p} \sim \exp \bar{U}t$, $p \gg 1$, i.e., the intermittency disappears from the leading order. Thus, the phenomenon of intermittency in a steady potential is associated with the presence of an infinite tail in its statistical distribution which seems to be unphysical.

More realistic is the development of intermittency in a potential (medium) that depends on time and is stationary only on the average. It is expected in this case that the solution ψ grows less rapidly, viz., as an exponential function of time because any maximal value of the potential persists for a finite period.

8.4. THE SHORT-CORRELATED RANDOM MEDIUM

In this section we turn our attention to the media which are stationary only on average. First, consider the ways of describing such media. Usually this description is accomplished in terms of the mean and r.m.s. characteristics of, e.g., dielectric permeability or — in magnetohydrodynamics — the mean helicity. A more detailed description assumes that the correlation tensor of fluctuations in a random medium is given or only some of its parameters are prescribed. For instance, when the Kolmogorov turbulence is mentioned, it is usually understood that the spectral index is given within the inertial range. Such description is insufficient for our purposes. When we are interested in the hardly probable states of the field that evolves in a random medium, we should

[c] The effective upper limit of the potential in quantum theory can be associated with Pauli's exclusion principle.

also be capable of describing the corresponding hardly probable states of the medium itself. This task is accomplished in the easiest way in the case of a medium which is extremely non-stationary. We speak of such media as short-correlated (in time).

To be specific, we describe a short-correlated potential U. This can be conveniently considered as a limiting case of a sequence of potentials $U^{\Delta}(t, x)$ which do not vary with time during periods of duration Δt: $(0, \Delta t)$, $(\Delta t, 2\Delta t)$, . . . , and are independent at different periods. In the limit $\Delta t \to 0$ we have

$$\langle (U(\mathbf{x}, t) - \langle U \rangle)(U(\mathbf{y}, t') - \langle U \rangle) \rangle = 2\delta(t - t')\, V(\mathbf{x}, \mathbf{y})\,. \qquad (9)$$

Following the normalization condition $(U^{\Delta})^2 \Delta t = 1$, for small Δt, the potential deviates from its mean value by approximately $(\Delta t)^{-1/2}$. Independence of the function $V(\mathbf{x}, \mathbf{y})$ on time implies that the potential is stationary on average.

The relation (9) describes our potential in terms of the correlation tensor. However, this is an incomplete description. Indeed, expression (9) allows us to evaluate the correlators of the type $\langle U(\mathbf{x}, t_1)\, U(\mathbf{x}, t_2) \ldots U(\mathbf{x}, t_n) \rangle$, defined at a single point, because under natural assumptions the short-correlated potential has the Gaussian distribution in time. In contrast, higher spatial correlations of the potential for separated points generally do not split into products of pair correlations and they cannot be reduced to (9). In more complicated situations, when the medium is not short-correlated, the Gaussian approximation cannot be applied to time-correlations as well (see, e.g., Molchanov *et al.*, (1985)).

This concept of short-correlated medium (i.e., short-correlated potential or other characteristics of the medium), which is so convenient for our purposes, is intrinsically connected with concepts of white noise and the Wiener process.

The Wiener (Brownian) random process is naturally associated with diffusion phenomena. However, here we also intend to use the Wiener process to specify potentials that would describe random interactions. The non-standard nature of this situation is underlined even by the adopted terminology: a mathematical concept that arises in the analysis of a trajectory of the Brownian particle now appears as, e.g., the velocity in the transfer equation! The point is that the properties of a random medium (e.g., the velocity), in which transfer, diffusion or self-excitation of a scalar (impurity) or vector (vorticity or magnetic field) occurs, enter the transfer equation as coefficients, just as the potential enters the Schrödinger equation. In this sense, our problem principally differs from Langevin's problem where the equation of motion includes a given random force (energy source). In our case the potential plays the role of a "driving belt" that connects energy sources with a self-exciting admixture through the characteristics of the medium.

The method of describing the short-correlated random process through the Wiener one can also be applied to the description of random media. We adopt the convention that at every position a particular Wiener process is given so that the whole set of these processes should be described as a function of t and $\mathbf{x}$: $W_t(\mathbf{x})$. Then the random short-correlated potential is given by

$$U(t, \mathbf{x}, \omega) = \sqrt{\sigma}\,\frac{dW_t(\mathbf{x})}{dt}.$$

The Wiener processes at different positions $\mathbf{x}$ and $\mathbf{y}$ can be correlated. A full description of the random medium demands that this correlation should be somehow characterized. Fortunately, in the majority of interesting cases, it is important that the correlation is strong only for nearby positions but it rapidly weakens with growing separation. This allows one to adopt a kind of discretization of space when it is considered that the properties of a random medium are uniform within the spatial correlation cells of certain correlation size while they are independent in different cells. In this case the Wiener process can be replaced by its finite-dimensional analogue, i.e., random wandering over a lattice. Sometimes, one is also forced to adopt time discretization.

The random media in which the property (9) is observed are usually called δ-correlated. Having in mind the further elucidation of this concept as described above, we adopt the notion of a short-correlated medium.

8.5. INTERMITTENCY IN NON-STATIONARY MEDIUM

Consider the simple case of an extremely non-stationary potential represented by a white noise in time with independent values for different spatial cells of a certain correlation scale:

$$U(t, \mathbf{x}, \omega) = \sqrt{\sigma}\frac{dW_t}{dt},$$

where W_t is the Wiener process. Adopting this form of generalized potential, we consider equation (6) as a finite-difference equation in time, i.e., it is presumed that the limit of infinitesimal correlation time of the potential is first taken and then the limit $\Delta t \to 0$ in (6). This order of the limits corresponds to the approach of Ito (1946). Otherwise, one can adopt the approach of Stratonovich (1966), which in this case corresponds to the reverse order of the indicated limit transitions. The conclusion about the presence of intermittency does not depend on the order of the adopted approach.

Equation (6) can be solved explicitly neglecting diffusion, i.e., for $\kappa = 0$, to give

$$\psi(\mathbf{x}, t, \omega) = \psi_0(\mathbf{x}) \exp\left(\sqrt{\sigma}\, W_t - \frac{\sigma t}{2}\right). \qquad (10)$$

The additional factor $\exp(-\sigma t/2)$ is due to the fact, as pointed out by Ito, that the correct differentiation of the Wiener process requires that its squared differential is also taken into account and that $(dW)^2 = dt$ (see, e.g., Gikhman and Skorokhod, 1965).

A typical value of the process W_t is of the order $t^{1/2}$ for large t, i.e., with the probability solution (10) which decays as $\exp(-\sigma t/2)$ at any point. However, at a small probability the Wiener process acquires values which are arbitrarily larger than $t^{1/2}$. Therefore, the solution undoubtedly possesses rare but intense peaks against the decaying background.

This can also be verified through the evaluation of high statistical moments of the solution. Suppose that the initial distribution $\psi_0(\mathbf{x})$ is independent of U or that $\psi_0(\mathbf{x})$ is a deterministic function. Applying the relation

$$\langle \exp(p\sqrt{\sigma} W_t) \rangle = \exp\left(p^2 \frac{\sigma t}{2}\right),$$

we obtain

$$\langle \psi^p \rangle = \langle \psi_0^p \rangle \exp\left[\frac{p(p-1)}{2} \sigma t\right],$$

i.e., the rate of exponential growth γ_p/p increases with the moment order as $\sigma(p-1)/2$. Thus, we see that a typical realization of the random field $\psi(\mathbf{x}, t)$ decays as $\exp(-\sigma t/2)$, the mean value ($p = 1$) always coincides with $\langle \psi_0 \rangle$, the mean square grows as $\exp(\sigma t/2)$, the fourth moment increases as $\exp(6\sigma t)$, etc.

This growth of moments is explained by the significant contribution of rare events. In other words, the complete set of realizations $\psi(\mathbf{x}, t, \omega)$ contains the realizations that grow at some positions. As can be seen from the form of (10) and the properties of the Wiener process, these rare functions are growing only for a finite time.

Intermittency leads to a very peculiar situation: even though ergodicity is preserved in intermittent media, practical application of the ergodic theorem to evaluate the averages over finite volumes is impossible. When considering a large but finite volume or a localized initial distribution $\psi_0(\mathbf{x})$, we encounter a very distinctive violation of ergodicity understood as coincidence of the averages over the statistical ensemble and spatial (sample) averages.

In contrast to statistical moments, in a finite volume V sample averages decay with unit probability:

$$\mu_p \equiv \frac{1}{V} \int_V \psi_0^p(\mathbf{x}) \exp\left(p\sqrt{\sigma} W_t(\mathbf{x}) - p\frac{\sigma t}{2}\right) d^3\mathbf{x} \underset{t \to \infty}{\propto} \exp\left(-p\frac{\sigma t}{2}\right).$$

Unlike statistical moments, the volume averages are random quantities themselves. Nevertheless, the rate of their decay $\gamma_p = -p\sigma/2$ is a deterministic quantity.

The difference between the sample and statistical moments can be understood in the following way. For a localized initial distribution (zero beyond certain radius) or in a finite volume, high peaks persist only for a finite period in the majority of realizations. Sooner or later, the characteristic separation of these peaks becomes greater than the size of the region occupied by the field and the peaks are present only in an exponentially small number of realizations. This low probability is sufficient for the peaks to make a significant contribution to the statistical average determined by all realizations. But the spatial average μ_p taken over a given realization evidently becomes insensitive to these peaks when they become too rare. The decay time of the number of peaks is of the order of $\tau_p \ln(V/r_0^3)$, where $\tau_p = 2/\sigma(p-1)$ is the characteristic time of growth of the p-th moment and r_0 is the correlation scale of the potential. Thus, for $t > \tau_p \ln(Vr_0^{-3})$, the given volume V ceases to encompass the peaks that considerably affect the p-th sample moment.

The intermittency survives even in the presence of diffusion, at least in the limit of small κ (Zeldovich *et al.*, 1987). Of course, the growth rates now become dependent on κ. For $\kappa \to 0$ the functions $\gamma_p(\kappa)$ remain continuous, therefore the values of $\gamma_p(\kappa)$ differ for different p for small κ. When κ grows, they vanish for certain values $\kappa_2, \kappa_3, \ldots$. It can be shown that $\gamma_1(\kappa) \equiv 0$, while $\gamma(\kappa) < 0$ for $\kappa < \kappa_0 < \kappa_2$ and $\gamma(\kappa) \equiv 0$ for $\kappa > \kappa_0$. The dependence of γ_p on κ for integer p can be obtained explicitly from the moment equations. However, the fact that for $p = 1$ the moment $\langle \psi \rangle$ is constant while a typical realization ψ decays for $\kappa < \kappa_0$ implies that all $\langle \psi^p \rangle$ for $p > 1$ grow exponentially at least for $\kappa < \kappa_0$. Thus, for $\kappa < \kappa_0$ the intermittency is no less pronounced than for $\kappa = 0$. For $\kappa > \kappa_0$, the intermittency can be noticed only in high statistical moments (see Fig. 8.1).

The discussed properties of the growth rates are also typical of the problem of transfer of a vector in a random medium, e.g., magnetic field in a turbulent flow of conducting fluid (turbulent hydromagnetic dynamo) (Molchanov *et al.*, 1985). The basic difference is that for sufficiently small (magnetic) diffusivity even a typical realization of a vector field can grow exponentially. In the case of a vector field, the peaks correspond to the structures of the type of magnetic ropes or layers.

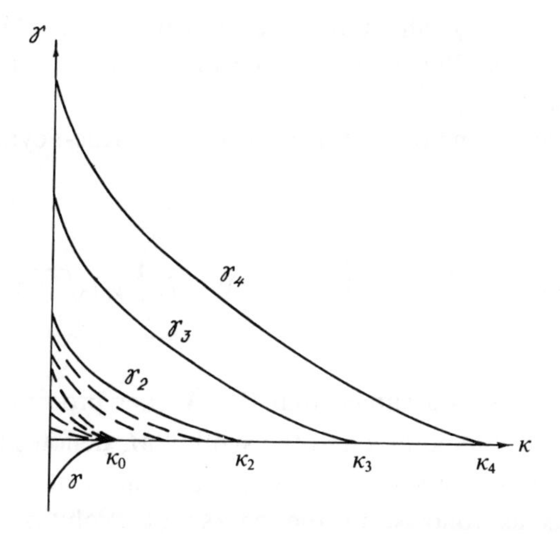

Fig. 8.1. Schematic dependence of the growth rate of a scalar field on diffusivity. The growth rates of the fractional moments for $1 < p < 2$ are shown by dashed lines.

8.6. STRUCTURE OF THE MOMENTS OF A SCALAR FIELD IN RANDOM MEDIUM

In order to get a deeper insight into the phenomenon of intermittency, it is useful to derive and analyze the equation for the moments of a random scalar field $\psi(\mathbf{x}, t)$ (Sokoloff and Shumkina, 1988). The term $U\psi$ in the equation

$$\frac{\partial \psi}{\partial t} = U\psi + \kappa \Delta \psi \tag{11}$$

that describes the evolution of ψ is the main obstacle in the derivation of these equations. Both multipliers, U and ψ, include randomness which enters ψ in a very complicated, nonlinear way, reflecting the history of evolution. This is why equations for averaged quantities can be derived more easily from the explicit solution of equation (11), rather than from the equation itself, because in the solution the random behaviour is expressed explicitly, albeit in a complicated form. The apparatus of Wiener integrals allows such a solution to be obtained even for an arbitrary function $U(\mathbf{x}, t)$.

The corresponding relation is known as the Kac-Feynman formula. It is given by

$$\psi(t, \mathbf{x}) = M_{\mathbf{x}}\left(\exp\left(\int_0^t U(\sqrt{2\kappa}\mathbf{W}_s)ds\right)\psi_0(\sqrt{2\kappa}\mathbf{W}_t)\right) \quad (12)$$

(cf. (7)), where it is assumed that the Wiener process $\mathbf{W}_t$ starts at the moment t from the point $\mathbf{x}$ and the symbol $M_{\mathbf{x}}$ denotes the average over such trajectories (the Wiener integral). The physical meaning of expression (12) is as follows: in the course of evolution, the field ψ is diffusionally transferred in space (the symbols $\mathbf{W}_t$ and $M_{\mathbf{x}}$ are responsible for this) and simultaneously grows at the rate U which varies from one position to another (the term $\exp(\int_0^t U ds)$). These heuristic arguments can be given in further detail but this is not quite necessary. Indeed a favourite peculiarity of any equation, including the differential ones, is that, after any suspicious heuristic arguments that lead to a certain solution, it can be validated by direct substitution into the equation and making sure that the equation turns into an identity. For the Cauchy type of differential equations, the solution is usually unique.

Let us do this with expression (12). Recall that the increments of the Wiener process are independent and we can consider that it starts afresh at any new moment. Therefore, (12) can be recast as

$$\psi(\mathbf{x}, t + \Delta t) = M_{\mathbf{x}}\left(\exp\left(\int_t^{t+\Delta t} U(\sqrt{2\kappa}\mathbf{W}_s)ds\right)\psi(\sqrt{2\kappa}\mathbf{W}_{\Delta t}, t)\right), \quad (13)$$

where the function ψ at the moment t is substituted for the initial distribution $\psi_0(\mathbf{x})$. Let us now evaluate the integrand in (13) with accuracy up to Δt taking into account that $\mathbf{W}_{\Delta t} = 0[(\Delta t)^{1/2}]$:

$$\exp\left(\int_t^{t+\Delta t} U(\sqrt{2\kappa}\mathbf{W}_s)ds\right)\psi(\sqrt{2\kappa}W_{\Delta t}, t)$$

$$= \psi(\mathbf{x}, t) + U(\mathbf{x})\psi(\mathbf{x}, t)\Delta t + \frac{\partial\psi}{\partial x_i}\sqrt{2\kappa}(\mathbf{W}_{\Delta t})_i + \kappa\frac{\partial^2\psi}{\partial x_i \partial x_j}(\mathbf{W}_{\Delta t})_i(\mathbf{W}_{\Delta t})_j .$$

Since we take the limit $\Delta t \to 0$, the higher-order terms can be neglected. Now the Wiener integral on the right-hand side can be easily evaluated and this approximation reduces the calculation of the correlators of the Wiener process. The result reads

$$\psi(\mathbf{x}, t + \Delta t) = \psi(\mathbf{x}, t) + U(\mathbf{x})\psi(\mathbf{x}, t)\Delta t + \kappa\Delta\psi\Delta t . \quad (14)$$

Transposing $\psi(\mathbf{x}, t)$ to the left-hand side of the equation, dividing the equation by Δt and taking the limit $\Delta t \to 0$, we recover equation (11).

The Kac-Feynman formula admits both deterministic and random potentials U. This solution (12) can be used for deriving the evolutionary equations for the statistical moments of the field ψ. For the sake of simplicity, we restrict ourselves to the equation of the first moment $\langle\psi(\mathbf{x}, t)\rangle$ for a short-correlated potential.

The idea behind this derivation is the following. Consider the short-correlated potential U as a limit, for $\delta \to 0$, of the potentials U_δ which are time-independent over the time intervals $[0; \delta)$, $[\delta; 2\delta)$, $[2\delta; 3\delta)$, ... separately and are statistically independent at different intervals. Consider also $\langle U_\delta\rangle = 0$. We recall that the potentials U_δ have typical values of the order of $\delta^{-1/2}$.

Let us calculate the increment $\psi[\mathbf{x}, (n + 1)\delta] - \psi(\mathbf{x}, n\delta)$ to the accuracy of the order of δ. This can be accomplished with the help of the Kac-Feynman formula, just as it was done in the derivation of expression (14) above. We should only take into account that now the potential U is of the order of $\delta^{-1/2}$ rather than unity. As a result, we obtain

$$\psi[\mathbf{x}, (n + 1)\delta] - \psi(\mathbf{x}, n\delta)$$

$$= \psi(\mathbf{x}, n\delta)\, U_\delta(\mathbf{x}, n\delta)\delta + U_\delta^2(\mathbf{x}, n\delta)\delta^2\psi(\mathbf{x}, n\delta) + \kappa\Delta\psi(\mathbf{x}, n\delta)\delta. \quad (15)$$

The random field $\psi(\mathbf{x}, n\delta)$ on the right-hand side of this expression was formed during the period $t < n\delta$ when the values of the potential U were independent of $U_\delta(\mathbf{x}, n\delta)$. Therefore, $\psi(\mathbf{x}, n\delta)$ and $U_\delta(\mathbf{x}, n\delta)$ are independent random fields and they can be averaged separately over the ensemble of realizations of U. This leads to

$$\langle\psi[\mathbf{x},(n+1)\delta]\rangle - \langle\psi(\mathbf{x},n\delta)\rangle = \kappa\Delta\langle\psi(\mathbf{x},n\delta)\rangle + B_\delta^2(\mathbf{x})\delta\langle\psi(\mathbf{x},n\delta)\rangle,$$

where $B_\delta^2(\mathbf{x})$ denotes the average $\langle U_\delta^2(\mathbf{x})\delta\rangle$. Transposing, dividing by δ and taking the limit of $\delta \to 0$, we obtain the following equation for the first moment:

$$\frac{\partial\langle\psi\rangle}{\partial t} = \kappa\Delta\psi + B^2\psi\,.$$

The equations for higher moments

$$m_p(\mathbf{x}_1,\ldots,\mathbf{x}_p,t) = \langle\psi(\mathbf{x}_1,t)\,\psi(\mathbf{x}_2,t)\ldots\psi(\mathbf{x}_p,t)\rangle$$

can be obtained in an analogous manner and they have the form

$$\frac{\partial m_p}{\partial t} = \kappa(\Delta_1 + \ldots + \Delta_p)m_p + \frac{1}{2}\sum_{i,j} V(\mathbf{x}_i,\mathbf{x}_j)m_p\,. \tag{16}$$

Here, Δ_i denotes the Laplace operator in $\mathbf{x}_i$ and $V(\mathbf{x}_i, \mathbf{x}_j)$ is the correlation function of the potential U defined by [cf. (9)]

$$\langle U(\mathbf{x}_i,t)dt,\, U(\mathbf{x}_j,t)dt\rangle = V(\mathbf{x}_i,\mathbf{x}_j)dt\,. \tag{17}$$

The differential dt enters twice on the left-hand side of (17) while it appears only once on the right-hand side because the potential U is a generalized function, viz., a white noise rather than an ordinary function. Note that only the binary correlations of the potential enter equation (16).

In the derivation of (16), we consider the renovation time δ to be short as compared to the characteristic time τ by which the Wiener process is normalized but δ to be large in comparison with dt. In other words, here we followed the Stratonovich approach. Meanwhile, in the discussion above, the Ito approach was used. When ψ is considered as bacterial density, the Stratonovich approach assumes that the bacteria reproduce themselves instantaneously at a characteristic time much shorter than the time of variation of the potential (the food supply). In the Ito approach, the maturing time of a bacterium exceeds the variation time of the potential. Comparing the two corresponding results, we see that frequent reforms destroy the population. One can verify that the moment equations for equation (11) of Ito's approach are the following:

$$\frac{\partial m_p}{\partial t} = \kappa(\Delta_1 + \ldots + \Delta_p) m_p + \sum_{i<j} V(\mathbf{x}_i, \mathbf{x}_j) m_p . \tag{18}$$

These equations differ from (16) by the absence of the term $V(\mathbf{x}_j, \mathbf{x}_j)$. In order to keep our presentation continuous, here we study equation (18). In Chapter 9, where more specific problems of magnetic field evolution in turbulent media are considered, we adopt a more adequate Stratonovich approach.

We consider the case of statistically homogeneous and isotropic potentials where V depends only on $r = |\mathbf{x}_i - \mathbf{x}_j|$. Then equation (18) is invariant with respect to the translation of the "center of mass" whose coordinate is $\mathbf{R} = 1/p\, \Sigma_{i=1}^{p} \mathbf{x}_i$. It is convenient to introduce new coordinates $\mathbf{y}_i = \mathbf{x}_i - \mathbf{R}$ with the origin at the "center of mass". We are searching for translationally invariant solutions of equation (18), i.e., the functions m_p that do not depend on $\mathbf{R}$.

Intermittency arises from small diffusivities κ when it is natural to employ a semi-classical approximation[d] for solving equation (18). The potential $\Sigma_{i<j} V(\mathbf{x}_i, \mathbf{x}_j)$ in equation (18) reaches its maximal value at positions $\mathbf{x}_1 = \mathbf{x}_2 = \mathbf{x}_3 = \ldots = \mathbf{x}_p$ (i.e., $\mathbf{y}_i = 0$) and becomes negligible when $r = |\mathbf{x}_i - \mathbf{x}_j|$ exceeds the correlation radius of the potential U. It is natural to assume that the most rapidly growing solution of (18), which is most interesting, is concentrated at small r. We can approximate the function $V(r)$ by a parabolic law

$$V(r) = a - br^2 \tag{19}$$

[d] We understood the semi-classical approximation in a modern broad sense (Maslov and Fedorjuk, 1981) that includes also the so-called oscillatory approximation.

for small r. Here a is the mean square of the potential U and $r_0 = (a/b)^{1/2}$ is the correlation radius of U. Of course, the approximation (19) is valid only for $r \lesssim r_0$ and when the function V does not include any dimensionless parameters considerably differing from unity (e.g., the hydrodynamic Reynolds number in the problem of magnetic field self-excitation).

Substituting the approximation (19) in equation (18) yields

$$\frac{\partial m_p}{\partial t} = \kappa \Delta m_p + \left[\frac{p(p-1)}{2} a - bp\rho^2\right] m_p , \qquad (20)$$

where Δ denotes the Laplace operator in $3p$-dimensional space of vectors y_i (with $i = 1, 2, \ldots p$) which are restricted by $\Sigma_{i=1}^{p} \mathbf{y}_i = 0$; $\rho = (\Sigma_{i=y}^{p} \mathbf{y}_i^2)^{1/2}$ is the distance from the origin in the $3p$-dimensional space.

Consider the solutions to equation (20) which preserve their shape, i.e., the eigenfunctions of the problems

$$\left[\gamma - \frac{p(p-1)}{2} a\right] \hat{m}_p = \kappa a \hat{m}_p - bp\rho^2 \hat{m}_p , \qquad (21)$$

where γ is the growth rate: $m_p(\mathbf{y}_1, \mathbf{y}_2, \ldots, \mathbf{y}_p, t) = \hat{m}_p(\mathbf{y}_1, \ldots, \mathbf{y}_p) \exp(\gamma t)$. Equation (21) is essentially the equation describing a harmonic oscillator in $3p$-dimensional space. For solutions decaying at infinity, the leading eigenvalue is given by

$$\gamma_0^{(p)} = \frac{p(p-1)}{2} a - 3p\sqrt{\kappa b p} \qquad (22)$$

and the corresponding eigenfunction is

$$\hat{m}_p^{(0)} = \exp\left\{-\frac{\rho^2}{2}\left(\frac{bp}{\kappa}\right)^{1/4}\right\}. \qquad (23)$$

This form is applicable up to the distance $\rho \sim r_0 p^{1/2}$. Beyond this radius, $\hat{m}_p$ falls off as $\exp(-A\rho)$ but the attenuating rate A depends on the detailed structure of V.

Following from (22), the intermittency is preserved only for sufficiently small values of κ. When considering the evolution laws of different moments, one should compare the values of the ratios $\gamma^{(p)}/p$. These ratios increase with p only for small κ. Using (22), we can also estimate the critical diffusivity $\kappa_{cr}^{(p)}$ defined so that for $\kappa < \kappa_{cr}^{(p)}$ the p-th moment is self-excited and for $\kappa = \kappa_{cr}^{(p)}$ the growth rate of the p-th moment, $\gamma^{(p)}(\kappa)$, vanishes. Equating $\gamma_0^p(\kappa)$ from (22) to zero we obtain

$$\kappa_{cr}^{(p)} = \frac{(p-1)^2 a^2}{36bp} \approx \frac{a^2}{36b} p \quad (p \gg 1) .$$

As a function of κ, the growth rate is continuous at $\kappa = 0$ but it depends on κ non-analytically, as $\gamma^{(p)} - \gamma^{(p)}(O) = O(\kappa)^{-1/2}$.

The eigenfunction (23) possesses the highest possible symmetry, the rotation group in $3(p-1)$-dimensional space. Note that only *a priori* the symmetry of the rotation groups in each three-dimensional space separately could be expected as $O(3) \times O(3) \times \ldots \times O(3)$. This property of maximal possible symmetrization can be encountered in other physical problems as well.

Semi-classical asymptotic forms for several higher eigenvalues and eigenfunctions of equation (18) can be obtained easily:

$$\gamma_N^{(p)} = \frac{p(p-1)}{2} a - 2\sqrt{\kappa p b}\left(N + \frac{3}{2} p\right), \quad N \ll \frac{1}{\sqrt{\kappa p b}}$$

while the eigenfunctions are obtained from (23) through multiplication by combinations of Hermite polynomials with sums of order equal to N.

The evolution of an arbitrary initial spatially uniform distribution of the p-th moment can be described with the help of an expression in the derived eigenfunctions. The solution is dominated by the leading eigensolution after a sufficiently long period.

Some useful information can also be obtained from the spectral properties, i.e., from the Fourier transforms of the eigenfunctions (2) given by

$$m_p(\mathbf{k}_1, \ldots, \mathbf{k}_p) = \left(\frac{bp}{\kappa}\right)^{3p/8} \exp\left\{-\frac{k^2}{4}\left(\frac{\kappa}{bp}\right)^{1/4}\right\} \delta(\mathbf{k}_1 + \ldots + \mathbf{k}_p),$$

where $k^2 = \Sigma_{i=1}^p \mathbf{k}_i^2$. This expression is applicable for $k > p^{-1/2} r_0^{-1}$.

Real potentials are always different from an idealized short-correlated potential. In particular, there can exist a limiting value $U_{\max}$ such that $|U| < U_{\max}$ with probability one. In this case, neither moment can grow at a rate exceeding $pU_{\max}$. Then the expressions obtained above are applicable only for $p < p_{\max}$ with

$$p_{\max} = \frac{2U_{\max}}{a} - 1 .$$

Notice that the eigenfunctions of the moment equations obtained do not contain any scales that considerably differ from unity. This probably implies that the intermittent distribution of the field consists of high peaks with approximately circular cross-sections formed by the lowest eigenfunction and connected by ridges associated with the higher eigenfunctions. Unlike this intermittency described by the simplest equation (11), in a more complicated example individual moments possess more versatile structure with many characteristic scales (see Chapter 9).

The solutions for equation (11) can be presented as

$$\psi \propto \exp\{\gamma(\kappa)t + \xi\sqrt{t}\},$$

where $\gamma(\kappa)$ is the rate of decay of a typical realization while ξ is a set of random quantities given at every position. For $\kappa \to 0$, the decay rate tends to $-1/2$ and ξ converges to a set of Gaussian random quantities uncorrelated among different positions. For non-vanishing diffusivity the random quantities ξ are essentially non-Gaussian. Indeed, for Gaussian ξ the moments' growth rates would be given by

$$\gamma_0^{(p)} = \gamma(\kappa) + \frac{p^2}{2} f(\kappa),$$

where f is defined as the dispersion of ξ.

Investigation of the asymptotic structure of the distribution of random quantities $\xi(\kappa)$ for small κ is the subject of further studies in relation to equation (11).

8.7. VORTICITY AND DEFORMATIONS IN A FLOW OF INCOMPRESSIBLE FLUID

We have discussed the description of intermittency in terms of higher statistical moments in detail. However, this language is still not commonly accepted in physics. This motivates our plan to illustrate in this section how physical insight can be gained from a knowledge of the behaviour of higher statistical moments, taking a simple hydrodynamic system as an example. In this intentionally chosen example, properties of lower (the first and second) moments are of minor importance.

As established by Bobylev's theorem (see, e.g., Deissler, 1984), the integral properties of the squared vorticity (enstrophy) and of the squared rate of strain are not independent. However, such identities are not available for higher powers of vorticity and strain. Such integrals are useful because they allow one to determine the characteristic of a flow (either vorticity or deformation) which forms a concentrated structure and which is distributed over the whole fluid volume. In other words, the integrals of higher powers of vorticity and deformation characterize the intermittency of a flow.

This section is aimed at establishing a correspondence between intermittency and integrals of vorticity and deformation. We emphasize that such integrals can be written out for any solenoidal vector field. We adopt a hydrodynamic terminology only in order to make the discussion more specific and to indicate possible applications. We do not rely on either equations of motion for a fluid or the Kelvin-Helmholtz theorem or any other particular property of fluids. Our results are also valid for the magnetic field (solenoidal axial vector field). In this case the considered integrals feature the electric current density and the tensor $1/2\,(\partial H_i/\partial x_j + \partial H_j/\partial x_i)$.

We recall that vorticity characterizes the rotation of a given fluid element. For a solid-body rotating with angular velocity $\boldsymbol{\omega}$, the linear velocity is $\mathbf{u} = \boldsymbol{\omega} \times \mathbf{r}$ and the vorticity is homogeneous, $\nabla \times \mathbf{u} = 2\boldsymbol{\omega}$. The vorticity is frozen into the ideal fluid in just the same sense as the magnetic field. In a two-dimensional case, this means only that the total (Lagrangian) derivative of vorticity vanishes. In a viscous fluid, the rate of strain tensor $D_{ij} = 1/2\,(\partial u_i/\partial x_j + \partial u_j/\partial x_i)$ determines the energy dissipation rate which is given by $-1/2\,\nu \int D_{ij}D_{ij}d^3_{\mathbf{x}}$ with ν as the fluid viscosity (the summation over repeating indices is understood).

Vorticity is a natural characteristic of an incompressible flow. To decide whether a given vorticity is weak or strong, it would seem natural to compare the mean squares of vorticity and the rate of strain (their mean values, or course, vanish for an isotropic random flow). However, Bodylev's theorem states that these two quantities are equal, i.e., for a given r.m.s. value of the rate of strain, all incompressible flows are vortical to the same degree. For a compressible fluid, the integral of the squared vorticity is always smaller than the integral of the squared rate of the strain tensor.

Before proceeding to the analysis of higher powers of vorticity and deformation, we first prove Bobylev's theorem for the case of differential rotation with $\omega = \omega(s^2)$, where s is the distance from the rotation axis that passes through the origin. This flow is two-dimensional and the integration is carried out over the plane $z = 0$. Introducing the notation

$$\boldsymbol{\Omega} = \nabla \times \mathbf{u}, \quad \delta^2 = D_{ij}D_{ij}, \quad A^{(2)} = \frac{1}{2}\Omega^2 - \delta^2 ,$$

where summation over repeating subscripts is understood, we have

$$\int A^{(2)} dS = 2\pi \int_0^\infty \frac{ds\omega^2}{ds^2}\, ds^2 = 2\pi u^2(\infty) .$$

Thus, the considered integral vanishes when the fluid at infinity (or boundary) is at rest. In the general case, to prove this identity it is sufficient to verify that $A^{(2)}$can be written as a divergence of some vector. Indeed,

$$\Omega^2 = \varepsilon_{ikl}\frac{\partial u_l}{\partial x_k}\varepsilon_{imn}\frac{\partial u_n}{\partial x_m} = \frac{\partial u_i}{\partial x_k}\frac{\partial u_i}{\partial x_k} - \frac{\partial u_i}{\partial x_k}\frac{\partial u_k}{\partial x_i} ,$$

where we have used

$$\varepsilon_{ikl}\varepsilon_{imn} = \delta_{km}\delta_{ln} - \delta_{kn}\delta_{lm}\,.$$

On the other hand,

$$\delta^2 = \frac{1}{4}\left(\frac{\partial u_i}{\partial x_k} + \frac{\partial u_k}{\partial x_i}\right)\left(\frac{\partial u_i}{\partial x_k} + \frac{\partial u_k}{\partial x_i}\right) = \frac{1}{2}\left(\frac{\partial u_i}{\partial x_k}\frac{\partial u_i}{\partial x_k} + \frac{\partial u_i}{\partial x_k}\frac{\partial u_k}{\partial x_i}\right).$$

Due to incompressibility,

$$\frac{\partial u_i}{\partial x_k}\frac{\partial u_k}{\partial x_i} = \frac{\partial}{\partial x_k}\left(u_i\frac{\partial u_k}{\partial x_i}\right),$$

which shows that the difference $1/2\,\Omega^2 - \delta^2$ is a divergence. For a compressible fluid, we can easily show that

$$\int A^{(2)} dr^3 = -\int (\nabla\cdot\mathbf{u})^2 d^3\mathbf{r} = -\int\left(\frac{\partial\rho}{\partial t}\right)^2 d^3\mathbf{r}\,.$$

Thus, the r.m.s. vorticity is maximal for an incompressible fluid.

Now we can verify that the integrals of higher powers of vorticity and deformation are not connected by any identities. For the fourth degree, we have

$$\int A^{(4)} d^3\mathrm{r} \equiv \int\left(\frac{1}{4}\,\Omega^4 - \delta^4\right) d^3\mathbf{r} = 2\int \frac{\partial u_i}{\partial x_j}\frac{\partial u_i}{\partial x_j}\frac{\partial u_l}{\partial x_k}\frac{\partial u_k}{\partial x_l} d^3\mathbf{r}\,.$$

This integral does not vanish in the general case. The integral of the difference of arbitrary $2p$-th powers, $\int A^{(2p)} d^3\mathbf{r} = \langle A^{(2p)}\rangle$ can also be written out.

The physical meaning of the quantity $\langle A^{(2p)}\rangle$ is the following. If this is positive for large p, then the positions where the velocity changes rapidly in space are the positions where the vorticity is predominantly concentrated; otherwise, these are the places where deformation concentrates. For instance, for a differential rotation with an angular velocity $\omega = \omega_0 \exp[-(r/r_0)^2]$ we have $\langle A^{(4)} \rangle = (3/16)\omega_0^4 r_0^3$. Therefore, for a system of individual vortices in an incompressible fluid, $\langle A^{(4)} \rangle$ and all other quantities $\langle A^{(2p)} \rangle$ are always positive. When large fluid volumes have only a moderate number of vortices formed on their boundaries, then the places where the velocity strongly changes are the positions of concentrations of the rate of strain. Indeed, on the boundaries of these volumes, the velocity field can be represented as $u_i = c_{ij} x_j$ with $c_{ii} = 0$, as required by incompressibility. At the places where a moving fluid volume moves apart the fluid or where the fluid closes up behind it, it can be easily shown that the r.m.s. rate of strain greatly exceeds the r.m.s. vorticity, provided the matrix c_{ij} is nearly diagonal. At the lateral surfaces of a moving volume, the flow is close to the Couette flow, $\mathbf{u} = (u_x(y), 0, 0)$, for which the squared vorticity and the rate of strain coincide at every position. Therefore, for flows around produced by bulk motions of large fluid volumes, $\langle A^{(2p)} \rangle < 0$.

The flows produced by bulk motions of large fluid volumes through the surrounding fluid are unstable and degenerate into systems of vortices (Zeldovich and Kolykhalov, 1982). This means that the first of the possibilities considered above is more typical of fluid mechanics: $\langle A^{(2p)} \rangle > 0$. Thus, if the initial flow is dominated by the straining motion ($\langle A^{(2p)} \rangle < 0$) it transforms so that $\langle A^{(2p)} \rangle$ changes sign. In principle, the signs of the moments $\langle A^{(2p)} \rangle$ can depend on both p and the Reynolds number. This occurs when the flow is hierarchical with some members of the hierarchy being dominated by the vortical motion and the others by the straining motion.

The considered characteristics of the vorticity distribution can also be applied to the analysis of turbulent flows. The simplest case is the homogeneous turbulence in an infinite space (in which all integrals are normalized to unit volume) which has ergodic properties so that the averages over space and ensemble coincide.

The behaviour of higher spatial moments of the squared vorticity can be made use of when deciding whether vorticity is concentrated into filaments or sheets in a given flow.

8.8. VORTEX FILAMENTS IN HYDRODYNAMICS

An extreme expression of the idea of intermittency is a vorticity concentrated within thin filaments so that its distribution is described by

$$\mathbf{\Omega} = \phi \mathbf{n} \delta_2[\mathbf{r} - \mathbf{r}_0(l, t)],$$

where $\mathbf{r}_0(l, t)$ is the function that describes the filament shape and position, l is the length measured along the filament, t is the time, δ_2 is the two-dimensional Dirac's delta-function in the planes orthogonal to the filament at its every point, and $\mathbf{n}$ is the unit vector tangent to the line $\mathbf{r} = \mathbf{r}_0(l, t)$. Obviously, both the planes on which δ_2 is defined and the vector $\mathbf{n}$ change their orientations from point to point and with time.

Isolated vortices are often introduced as a simplification for deriving the finite-difference analogue of hydrodynamic equations in partial derivatives. Hydrodynamics of discrete vortices is considered in detail by Gledzer *et al.*, (1981).

Agishtein and Migdal (1987) propose to use the discretized arrays of vortex filaments as a basis for the description of developed turbulence. We emphasize that only in intermittent situations can individual vortex filaments be considered as real objects. Unlike the vortices considered by Gledzer *et al.*, these vortices are real physical objects rather than the mesh-dependent idealizations of a numerical model.

Following from the solenoidality of vorticity, $\nabla \mathbf{\Omega} = 0$, the factor ϕ, the circulation ($\phi = \int \Omega \, dS$), must be constant along the filament. Moreover, in a given filament, ϕ is independent of time in an inviscid fluid.

Note that such vortex filaments can arise spontaneously only in three-dimensional flows because in two dimensions, the velocity field $\mathbf{v} = (v_x, v_y)$ of an incompressible fluid is expressed through a single scalar function and concentration in spots becomes more probable.

The approximation of isolated delta-functional vortex filaments is adequate for describing the global motion of filaments and fluid but is inadequate for describing viscous energy dissipation. Viscosity determines the true thickness of the filaments.

The idea of turbulent flow that consists of individual vortex filaments obviously represents a crude approximation to reality. The deviation from Kolmogorov's laws indicate a fractal character of the system of vortex filaments (see Chapters 5 and 7). Even an individual filament can be fractal. But a hierarchical picture of combining smaller vortices into isolated "intermittent", larger vortices can also have global fractal properties.

Concentration of a vector field into filaments is also typical of other physical fields. The next chapter is devoted partially to magnetic ropes in a moving plasma. These ropes are conceptually similar to the magnetic tubes of Michael Faraday. In our century, a hundred years after Faraday, individual vortex filaments are revealed in reality in rotating superfluid liquid helium. The intensity of every filament, determined by the quantum condition, is proportional to Planck's constant. It can already be seen from this fact that this intensity does not depend on time in any motion of the liquid.

The mutual interaction of ropes formed by a physical vector field depends on the nature of the field: parallel vortex filaments repel each other, while neighbouring magnetic ropes do not interact at all because the magnetic field vanishes outside a rope.

There are some indications that in turbulent flows, vorticity turns out to be concentrated into tubes (or in one-dimensional filaments in the approximation described above) rather than the velocity itself (Siggia, 1985; Hussain, 1986). One of the possible explanations of this is based on the fact that the streamlines are not translationally invariant while the vortex lines are translationally invariant. However, there is a more substantial and physical argument. A thin submerged jet of fluid with a given flux is unstable, while an isolated straight vortex line possesses a certain elasticity and is stable.

A similar situation occurs in magnetohydrodynamics: what concentrates within tubes is the magnetic field (and the current is flowing azimuthally) rather than the current.

And finally, note one more difference between the behaviours of hydrodynamic and magnetohydrodynamic vector fields. The hydrodynamic problem is closed and nonlinear. The instantaneous vorticity field completely determines the velocity field of an incompressible fluid when vorticity is both a smooth function of position and an array of δ-lines.

In a linear hydromagnetic problem, the behaviour of the magnetic field is considered against the background of independently prescribed velocity field. The existence of velocity fields with vortex filaments gives rise to interesting problems as regards the behaviour of the magnetic field in such flow. Will magnetic ropes arise in this velocity field? Will the ropes coincide with the vortex lines?

8.9. INTERMITTENCY IN NONLINEAR MEDIA

In this section we consider how self-excitation and intermittency are affected by the nonlinearity which is associated with the back action of the growing field on the medium. We stress that we consider here only the simplest forms of nonlinearity which cannot lead to formation of structures by themselves without the presence of randomness. It is natural to expect that nonlinearity leads mainly to suppression of the growth of high maxima of the field, thereby smoothing intermittency. However, the solution of this problem turns out to be less evident and essentially depends on both the temporal dependence of the potential and the form of the nonlinearity (Zeldovich *et al.*, 1987).

Here we restrict ourselves to the analysis of the simplest characteristic of the random intermittent field ψ, its limiting (for $t \to \infty$) probability distribution $\pi(\psi)$.

For the sake of simplicity, we put $\kappa = 0$, i.e., neglect field diffusion. Those results which are expected to be qualitatively dependent on the presence of weak diffusion are indicated below.

We consider two types of potentials. The first is the short-correlated (in time) potential. The second is a steady random potential $U = U(x, \omega)$. The potentials are always considered statistically as stationary, homogeneous and isotropic.

We begin our discussion with the nonlinearity that does not involve the potential,

$$\frac{\partial \psi}{\partial t} = U\psi - \varepsilon f(\psi)\,, \tag{24}$$

where ε is a parameter, f is the given nonlinear function, which grows sufficiently rapidly with ψ.

A random process described by equation (24) with the short-correlated potential,

$$U(\mathbf{x}, t, \omega) = \bar{U} + \frac{dW_t}{dt}$$

can be considered as one-dimensional diffusion along a line with ψ being the coordinate measured along this line:

$$d\psi = \psi dW_t + (\bar{U}\psi - \varepsilon f)dt .$$

The generating operator of this diffusion has the form

$$\hat{L} = \frac{\psi^2}{2}\frac{d^2}{d\psi^2} + (\bar{U}\psi - \varepsilon f)\frac{d}{d\psi} . \tag{25}$$

Thus, $1/2\,\psi^2$ plays the role of diffusivity while $-(\bar{U}\psi - \varepsilon f)$ is the advection velocity field. Denote by $p(t, \tilde{\psi}, \psi)$ the fundamental solution of the evolutionary problem corresponding to the operator (25) such that $p(0, \tilde{\psi}, \psi) = \delta(\tilde{\psi} - \psi)$ and

$$\frac{\partial p}{\partial t} = \hat{L}(\tilde{\psi})p = \hat{L}^*(\psi)p ,$$

where $\hat{L}^*$ denotes the operator adjoint to (25).

Suppose that there exists a limiting stationary density distribution $\pi(\psi) = \lim_{t\to\infty} p(t, \tilde{\psi}, \psi)$. This obeys the equation

$$\hat{L}^*\,\pi(\psi) = 0 \tag{26}$$

with the sole restriction of normalization $\int_0^\infty \pi d\psi = 1$. Equation (26) always has solutions. Indeed, the equation

$$\frac{1}{2}\frac{d^2}{d\psi^2}(\pi\psi^2) - \frac{d}{d\psi}\pi(\bar{U}\psi - \varepsilon f) = 0$$

has first integral

$$\frac{1}{2}\frac{d}{d\psi}(\pi\psi^2) = \frac{\bar{U}\psi - \varepsilon f}{\psi^2} + c_1 .$$

Solutions with $c_1 \neq 0$ have singularity at $\psi = 0$ and are therefore excluded. We see then that if the function f grows at least as fast as ψ^2 then a limiting distribution exists that is given by

$$\pi(\psi) = \frac{c}{\psi^2}\exp\left(2\int\frac{\bar{U}\psi - \varepsilon f(\psi)}{\psi^2}\,d\psi\right)$$

where the constant c follows from the normalization condition.

Consider $f = \psi^3$ as an example. Then

$$\pi(\psi) = \frac{(2\varepsilon)^{\bar{U}-1/2}}{\Gamma(2\bar{u}-1)}\psi^{2(\bar{U}-1)}\exp(-\varepsilon\psi^2), \tag{27}$$

where Γ denotes Euler's gamma-function. This yields an important condition $\bar{U} > 1/2$ for the existence of the density $\pi(\psi)$. For $\bar{U} < 1/2$, the density π concentrates at $\psi = 0$ and the solution decays. For $1/2 < \bar{U} < 1$, the function $\pi(\psi)$ monotonically decreases with ψ. In the limit $\varepsilon \to 0$ that corresponds to transition to the linear problem, we have $\pi \to 0$. In this limiting case, for $\bar{U} > 1$ the maximal value of the density decreases as $\varepsilon^{1/2}$. It can be shown that for non-zero diffusivity, $\kappa \neq 0$, the density $\pi(\psi)$ and all its derivatives vanish at $\psi = 0$. In a statistically homogeneous medium without diffusion, the solution eventually becomes identically distributed with density $\pi(\psi)$ at all positions with values of ψ being statistically independent at different points. The influence of diffusion leads to the appearance of spatial correlations which decay with distance at a rate that grows with the decrease of κ.

Statistical moments of the scalar field are given by

$$\langle \psi^p \rangle = \frac{\Gamma(p + 2\bar{U} - 1)}{2^{p/2}\Gamma(2\bar{U} - 1)} \varepsilon^{-p/2} .$$

Note that for $\bar{U} = 1$ distribution (27) coincides with the truncated Gaussian distribution since here $\psi \geqslant 0$.

The distribution obtained contains maxima of any height, but in the limit of small diffusivity the situation resembles the Gaussian one: the maximal field ψ very rarely shows strong deviation from the most probable value. These maximal values determine the magnitude of the mean quantities. Some traces of intermittency can be noticed in the fact that the probability density $\pi(\psi)$ only slowly decreases toward small values of ψ. This implies that the spatial distribution of the field ψ is characterized by regions where ψ is anomalously small.

For strong diffusion the field ψ decays. More interesting is the case of intermediate diffusivity which, when combined with nonlinearity, leads to an unusual distribution of the field: small values of ψ are most probable, but the mean quantities are determined not by these most probable values but, rather, by the relatively widely by spaced peaks of the maximal height allowed by nonlinearity. In Chapter 9 we will consider this intermediate intermittency in more detail.

Consider now a steady random potential, e.g., one with Gaussian statistical properties, mean value $\bar{U}$ and r.m.s. deviation σ. Obviously, for $f = \psi^3$ and in those regions where the potential is positive, the solution can be steady. Putting $\partial\psi/\partial t = 0$ in equation (24), we see that the solution ψ is distributed as $(U/\varepsilon)^{1/2}$. In the regions with $U < 0$, the solution tends to zero. For a Gaussian potential, the limiting distribution in the absence of diffusion is given by

$$\pi(\psi) = \frac{1}{2}\delta(\psi) + \frac{\varepsilon\psi}{\sqrt{\pi\sigma}} \exp\left(-\frac{\varepsilon^2\psi^4}{2\sigma}\right), \tag{28}$$

where we have put $\bar{U} = 0$ for simplicity. This distribution has two pronounced maxima, at the origin and for $\psi = \psi_* = (\sigma/2\varepsilon)^{1/4}$. Thus, we have an island intermittency: half of the space is occupied by the field distributed around the maximum ψ_* while the remaining half is empty. When the mean value of the potential differs from zero, $\bar{U} \neq 0$, the ratio of the volumes of filled and empty regions becomes different and the problem of percolation properties of those regions arises, similar to that considered in Chapter 6. Weak diffusion smears the delta-function maximum of the distribution (28) but this does not affect its intermittent character.

Consider now the case where nonlinearity enters the potential,

$$\frac{\partial \psi}{\partial t} = g(\psi)\, U(\mathbf{x}, t, \omega)\psi\,.$$

For the simplest case of step-function nonlinearity, $g = 1$ for $\psi < \psi_m$ and $g = 0$ for $\psi \geqslant \psi_m$. Let us first discuss the short-correlated potential. If $\bar{U} > 1/2$, the solution exponentially grows up to the level ψ_m and then remains constant. The case $\bar{U} < 1/2$ is less simple. At the linear stage, a typical realization of the field decays exponentially but high moments grow, i.e., sparse peaks are present. However, every realization is present at positions where $U > 1/2$ and ψ grows locally. Those maxima of the solution that exceed ψ_m in the course of this growth remain at this high level ψ_m forever. The resulting stationary picture is the following: at some widely scattered positions, there are high maxima of the height ψ_m, while between them the field decays exponentially. In the considered simple example, all peaks are identical like randomly scattered telegraph poles fixed at their positions. In order to diversify this picture, it is sufficient to consider $g(\psi)$ that smoothly decreases at infinity. Consider, e.g., $g = (1 + \psi/\psi_m)^{-1}$. For simplicity put $\bar{U} = 0$. Then

$$\psi(x, t, \omega) = \begin{cases} \psi_0 \exp\left(W_t - \dfrac{t}{2}\right), & \psi < \psi_m \\ \psi_m W_t, & \psi \gtrsim \psi_m \end{cases}.$$

We can see that at the places where the solution has not reached the level ψ_m, it exponentially decreases with time. This exponential decay does not occur in sparse peaks whose height exceeds ψ_m owing to the growth of the linear stage. These peaks can survive for a long time, but the one-dimensional Wiener process is recursive and therefore the field at the peaks eventually decreases down to ψ_m, and further. Of course, oscillations around the level ψ_m are also possible but their probability is lower the longer the period of oscillation of the solution. As a result, a typical realization decays. However, against this fading background stand the peaks which determine the mean values, i.e., the intermittency persists.

Indeed, even though a typical realization decays, the mean value $\langle\psi\rangle$ is independent of time. Consider the situation where at a certain moment t_m the solution intersects the level ψ_m, say it becomes equal to $2\psi_m$. Since W_t is a random process with independent increments, we can consider the level $\psi = 2\psi_m$ as the starting point. Then the probability P that up to the moment t the trajectory will experience one more improbable event and fall down to $\psi = 1\,\psi_m$ decreases with time, similar to the probability of remiss in can tossing (see Chapter 2), and is proportional to $(t - t_m)^{-1/2}$ Since a typical realization of W_t is also of the order $t^{1/2}$, the average value given by $\langle\psi\rangle = P(\psi)\psi = P\psi_m W_{t-t_m}$ is constant in time. This fact is true for any dependence of g on ψ, following from the vanishing of the mean value of Ito's integral $\int g(\psi)\psi(W_t)dW_t$ (McKean, 1969).

Higher statistical moments grow as $\langle\psi^p\rangle \propto t^{-1/2}t^{p/2} \propto W_t \propto t^{(p-1)/2}$. It can be expected that this weak power-law growth of the moments is stabilized by weak diffusion. Thus, for $\kappa \neq 0$ the peaks that arise are distributed over their heights with the Gaussian-type distribution. However, the field distribution in general is far from being Gaussian. Between the widely spaced peaks (whose height is inversely proportional to κ raised to a certain power), the field is weak and they are determined by their diffusion from the peaks. Meanwhile, in a field with Gaussian distribution, a typical realization behaves like the mean value, with deviations of the order of the r.m.s. value. Of course, for sufficiently large κ the diffusion smears the peaks and intermittency disappears.

In the case of a steady potential and step-function form for $g(\psi)$, considering again the regions with positive and negative potentials separately, we obtain the following limiting distribution of the type (28):

$$\pi(\psi) = p\delta(\psi) + (1 - P)\delta(\psi - \psi_m)\,,$$

where P is the probability of the potential having negative values. For more complicated forms of $g(\psi)$ the limiting distribution exists only when diffusivity is non-vanishing. This limiting distribution also possesses two maxima which already are not delta-functions while the quantity similar to ψ_m is now determined by κ.

Thus, for a steady potential, the island-type intermittency occurs independently of the character of the local nonlinear interaction.

The intermittency is even more pronounced when the nonlinearity is non-local, for instance when the nonlinear function in (24) depends on $\langle\psi\rangle$ or $\langle\psi^p\rangle$ rather than on ψ. It is clear that in this case $\langle\psi^{p_0}\rangle$ becomes independent of time, instead of $\langle\psi\rangle$. All moments of orders $p < p_0$ with some p_0 dependent on the form of the nonlinearity and a typical realization of the field decay exponentially, while moments with $p > p_0$ continue their intermittent growth typical of the linear problem.

Consequently, the character of the nonlinear solution essentially depends on the temporal behaviour of the potential and on the form of the nonlinearity. In applications, additional restrictions of conservation of a certain moment is often imposed, e.g., the mean value in the problem of bacteria regeneration with restricted food supply (Zeldovich, 1983) or the mean square (the energy) in the problem of radio wave propagation in a random medium (Rytov *et al.*, 1978). Nonlinearity can be of a more complicated nature, e.g., combining deterministic nonlinear damping and nonlinear modification of the random potential. In this case, the properties of the nonlinear solution are determined by the competition of the two nonlinear effects with one smearing out intermittency and the other enhancing it. Detailed quantitative analysis of this competition is a subject of further research.

8.10. PROPAGATION OF FIELD FRONTS IN RANDOM MEDIA

As a rule, the discussions above presumed that initially, the field is distributed over the whole space, $\psi_0(\mathbf{x})$. This could either be a uniform distribution $\psi_0(\mathbf{x}) = \text{const.}$ or a spatially uniform random field. When the initial field $\psi_0(\mathbf{x})$ is concentrated within a finite volume, at later stages it can not only grow at a given point but also propagate due to diffusion.

It is well known that a passive admixture diffusively propagates according to the law $x \propto (\kappa t)^{1/2}$, i.e., the propagation velocity decreases with time as $v \propto t^{-1/2}$. When diffusion is combined with local generation, the propagation velocity becomes constant in time. This process of propagation of the admixture front in an unstable medium is the subject of the theory of Kolmogorov, Petrovsky and Piskunov (1937); these ideas are widely applied in the combustion theory (see Zeldovich *et al.*, 1985). Let us illustrate this problem by the following simple example of deterministic potential. Consider the equation

$$\frac{\partial \psi}{\partial t} = U_0 \psi + \kappa \Delta \psi, \quad U_0 = \text{const.}\,,$$

with the initial distribution concentrated at the origin. The solution is evidently given by

$$\psi(\mathbf{x}, t) = \frac{1}{(4\pi\kappa t)^{3/2}} \exp\left(U_0 t - \frac{r^2}{4\kappa t} \right), \quad r = |\mathbf{x}|\,.$$

The argument of the exponential function can be recast as

$$U_0 t - \frac{r^2}{4\kappa t} = -\frac{1}{4\kappa t}(r - 2\sqrt{U_0 \kappa t}\,)(r + 2\sqrt{U_0 \kappa t}\,),$$

from which it follows that the surface ψ = const. propagates with a velocity

$$v = 2\sqrt{\kappa U_0}\,.$$

In a random medium, a typical realization propagates with the velocity

$$v = 2\sqrt{\kappa\left(\bar{U} - \frac{\sigma^2 \tau}{2}\right)}\,.$$

Simultaneously, the mean values also propagate so that the p-th moment's propagation velocity is given by

$$v_p = 2\sqrt{\gamma_p \kappa}\ ,$$

where γ_p is its growth rate. In a short-correlated medium, in the limit of weak diffusion we have

$$\gamma_p = \bar{U} + \frac{(p-1)\sigma^2\tau}{2}\ .$$

Consequently,

$$v < v_1 < v_2 < \ldots\ .$$

This means that the boundary of the distribution of ψ is very indented and fractal. The propagation occurs by the protrusion of the leader tongues.

When analyzing the propagation of the field ψ, it is convenient to interpret ψ as the bacteria number density, the potential U as the rate of their reproduction or death, and the diffusivity κ as the rate of their spreading in space. The field ψ can be visualized as a map of the propagation of an epidemic disease caused by the bacteria. Then the fractal properties of the distribution boundary suggests not only a slow propagation of the infection from one neighbour to another but also that improbable but catastrophic intrusions of the infection into unaffected regions are important in spreading an epidemic.

At first sight, it would seem that nonlinearity is of minor importance for the propagation of the field ψ, because near the front the values of ψ are usually small and the nonlinearity seems to be negligible. However, the very possibility of propagation of the statistical moments at velocities exceeding that of a typical realization is associated with the existence of peaks whose occurrence has low probability but the height is large; with these peaks nonlinearity is essential. As a result, in nonlinear problems all mean quantities propagate at the same velocity but the propagation front has a finite thickness. In the case of strong diffusion this fractal zone consists mainly of sparse peaks. Returning to the analogy of an epidemic disease, we notice that this is the way in which epidemics spread in our mobile age. Adequate sanitary and medical measures (the nonlinearity) result in a relatively low number of seriously affected persons. In contrast, for relatively low mobility, high reproduction rate and weak nonlinearity, the propagation front consists mainly of phases with high values of ψ and only rare holes. This picture seems to be adequate for describing catastrophic epidemics of the Middle Ages when only the strongest persons survived them. Of course, such comparisons are reasonable only for similar deseases which were fought with similar efforts by physicians: influenza and plague spread differently.

We would like to illustrate the process of propagation of an epidemic in a random medium using the communication of such a thoughtful observer as Gregory, Bishop Turonesis, the author of "Historiarum libri X" (6th century AD). Describing the plague epidemic of the year 588, he noted (Gregorii, Hist., IX, 22) that the infection was conveyed by a ship which came to Marseille from Spain. Since the inhabitants of the town bought goods from the ship, very soon one of the houses with eight inhabitants had perished. However, the epidemic did not spread immediately after this. Only after some time were all parts of the town infected. Meanwhile, the bishop and the remaining people escaped to St. Victor Basilica. The disaster ended after only two months, but when people came back to their homes the epidemic flared up again. The author noted that such reccurrences of the plague repeated many times before the final end of the epidemic. From the viewpoint of theory, the role of the ship corresponds to a "quasioptimal" trajectory along which the solution propagates (see equation (8)). The first event of the infection corresponds to the predecessors in the form of leading tongues followed by a universal infection and the formation of a local minimum (people escaped to the basilica). The following events are essentially nonlinear and they go beyond the scope of the present theory. The end of epidemics due to "shortage of fuel" and the relapses that followed are considered, e.g., by Polak and Mikhailov (1983). Notice that Gregory compares the spread of the epidemic with the spreading of fires. Although commentators note that this comparison is inspired by "Aeneis" of P. Virgilius Maro (II, 304), this image agrees perfectly well with modern physical views.

8.11. THE INTERMITTENT WORLD

To conclude this chapter, we would like to discuss the role of intermittency in phenomena which are relatively far from the interests of a physicist. In contrast to the problem of intermittency of magnetic fields in a moving conducting medium (see Chapter 9) where the analysis is based on for laws of physics and is essentially quantitative, for the case considered here only a qualitative description is as yet possible. However, the examples of this section will make clear that intermittency is a very widespread phenomenon which is far from restricted to magnetohydrodynamics in particular or physics in general. (At any rate, the importance and beauty of a theory are not directly connected with the range of its applicability.) In every day life as well as in the Arts, we constantly encounter situations which hardly resemble the artificially regularized conditions that are so typical of a physical theory and in which the role of chance is intentionally smoothed out.

We can come across intermittency every day indeed. Imagine, if you can, that you are walking around in the renowned Murom woods, the birthplace of the characters of Russian folklore. It is early summer and the mushrooms have just appeared. An inexperienced mushroom collector returns home with an empty basket after having combed the woods. His confusion is even stronger because he is sure that somewhere in the woods there must be a narrow strip studded with mushrooms. This is a typical manifestation of intermittency at a linear stage of an instability. Of course, this phenomenon can be considered from another standpoint. The strip rich with mushrooms surely has a special type of vegetation, temperature and illumination which altogether explain an abundance of early mushrooms. An exact determination of all such conditions would allow one, in principle, to find other places which can make any mushroom fanatic happy. This is a more common formulation of the problem. Nevertheless, the multitude of equally important factors and their random character result in an intermittent character of the distribution of mushroom-rich places in a forest and their productivity. Thus, at least equally valuable is another formulation of the problem, which considers the abundance of mushroom as a random field.

During the early mushroom season, their distribution is very uneven. After a few weeks mushrooms can be found in almost every part of a forest. Even though the gradients of mushroom productivity become less pronounced, your good friend would always show you a special place where mushrooms remain exceptionally numerous even at this "non-linear" stage. Similar notions are characteristic of many other realms of our life: artists rely on inspiration, gamblers wait for the rare lucky coincidences, sailors are terrified by "the ninth wave".

In contrast to the woods, intermittency at a farmer's fields is smoothed out by the cultivation of the land, care of the crops, etc. Here intermittency can be noticed on another level and this can be revealed when one compares the effectiveness of different farms and considers some statistical information. This intermittency is typical of many human activities. In a jocular form, this is expressed by the observation that "80 percent of beer is consumed by 20 percent of the population". It is difficult to judge how stable this ratio 20:80 is but this observation is undoubtedly well founded. The uneven distribution of welfare provides another example. These effects can be illustrated by observations of student groups formed on some competitive basis. Even when competition is strong and the teacher expects to have a uniformly strong class, very soon two small groups composed of very strong and very weak students separate from the moderate majority. These two groups essentially determine the progress of the whole class. The initial level of the applicants determines the average level of the class but this structure with a few leading and a few backward students is very stable. Only special measures like a reduction in requirements to the applicants or collection of backward students in special classes can suppress such a structure. As shown by our experience, such attempts lead to catastrophic consequences for the students.

When thinking over the results concerning intermittency, one cannot help switching one's thoughts to social analogies. We have interpreted the equation $\partial\psi/\partial t = U\psi + \kappa\Delta\psi$ used for our study of intermittency above as a model of the evolution of bacterial population in a medium with randomly distributed food supply and other living factors when the population is able to spread diffusionally in space. This model predicts that the bacteria gather in comparatively stable clusters at places where the environment is most favourable for their development. This result can be verified by means of special microbiological investigations. But, evidently, this picture is surprisingly a good description of the distribution of human population. There are many levels of concentration of people, from villages to towns, cities and such monstrous cities as Tokyo, New York or Moscow. If we exclude such unique cases as Singapore, where this book is published, cities occupy a comparatively small territory while the population density within them is incomparable with those outside these structures. To be specific, consider two examples. In Russia (it is essential here that this geographical unit does not coincide with the USSR) population living outside the small or large settlements is practically absent. This kind of population is typical of another part of the USSR, Lithuania, where "Viehkiemiai" or small isolated farmer settlements are rather numerous. However, even in Lithuania, several million people live within a 10^5 km^2 area, while in Vilnius, the capital of Lithuania, the population of 0.5 million lives within a circle of 10 km diameter. Tens of smaller towns in Lithuania occupy a negligible part of the territory but they give shelter to the majority of the population. The completely unoccupied territories (woods, swamps, etc.) are not very vast; their area compares with that of the farmer settlements. Thus, we show evidence of two types of intermittency: in Russia the population density is negligibly small in a typical realization and almost the total population is concentrated in an intermittent structure while in Lithuania the intermittency peaks stand against a background of a typical realization significantly differing from zero. Different kinds of intermittency are present within the cities as well. For buildings, in Vilnius typical large districts have private houses (the typical realization) while in Moscow people live mainly in large apartment blocks (intermittent structure). Moreover, lower levels of intermittency (districts uniformly and densely covered with buildings) are permanently supplanted by higher levels (isolated huge apartment blocks). This growth tendency of higher elements of an intermittent hierarchy is typical of Russia where the inhabitants of rural areas strive to move to cities and from smaller cities to larger ones. Within cities, there is a tendency of replacing the traditional uniform buildings of mixed apartment houses, offices, culture and amusement centers with a non-uniform distribution where the inhabitants live in one district, have jobs in another and amuse themselves in a third. This tendency of declining the diverse types of housing in favour of the highest level of the intermittent hierarchy is one of the major social problems of this country.

Surprisingly enough, this phenomenon even allows us to trace the differences of Ito's and Stratonovich's approaches to the evolutionary equations. When unreasonable and unpredictable political novelties occur so frequently that people cannot get adapted to them (Stratonovich-type regime is replaced by the Ito-type), this results in degeneracy of the intermittent hierarchy (the population moves to cities from the rural areas).

Another urgent social problem is bureaucracy. Here, one is tempted again to cite analogies with the simple results of the intermittency theory. We already know that random media are most suitable for the development of intermittency. Having compared the hierarchical elements of this structure with bureaucratic clans, we can understand why the stagnation periods in social development are marked by the catastrophic growth of bureaucratic mafia groups even if initially the society was dominated by honest people. The frequent shaking up of the bureaucratic system can slow down this growth and even suppress the formation of small and weak groups. But they can do nothing against the inevitable development of the most influential groups of bureaucrats. The only way to supress their growth consists in removing the mechanism of instability by introducing some social antibureaucratic mechanisms.

Of course, similar analogies are not restricted to sociology. Let us mention an example from evolutionary biology. We can consider the coordinate x in our simplest equation as the coordinate in the "space of properties" of organisms and the potential as a measure of the ability of an organism to survive and reproduce which depends on its particular properties. Then the formation of the structures can be interpreted as the formation of the species at the early stages of evolution. At subsequent stages, the species are fixed by various biological barriers that prevent the exchange of genetic information between different species (an analogy of nonlinear diffusion that promotes the development of the structures). The dimension of such space is unclear but the qualitative properties of intermittency are practically independent of the dimension.

The imagination can suggest more examples of this kind. We think that this implies that the concept of intermittency reflects something essential in the laws of nature and society. However, one should not be so excessively enthusiastic as to drag all the known structures up to the simplest results concerning intermittency which were discussed in this book. The systems considered in this section are much more complicated than a bacterial population in an inhomogeneous medium. Even if in some aspects our life hardly differs from the evolution of bacterial associations, in other aspects they are obviously different. In particular, the progressive growth of cities is undoubtedly connected with the more or less deliberate settling of people at places where life is more attractive. The effect of free will is, of course, neglected by our simplistic equation. The birthrate of the population of the largest cities cannot account for the total birthrate, and the growth time of cities is not directly connected with the activity period of a given generation. The similarity in manifesting different kinds of intermittency is associated with a weak dependence of the basic features of intermittency on a specific form of evolutionary equation. This does not suggest that the simplest equation of the bacterial evolution can be applied to sociology.

This universality of intermittency leads to an important mathematical problem, that of distinguishing and describing the classes of evolutionary operators with random coefficients which lead to intermittent phenomena. When properties of intermittent structures cease to be universal, general considerations are of little help. Let us turn to an example where this feature is essential.

The conglomerations of bacteria considered above have an approximately circular shape. Due to the solenoidality of the field, filaments arise in the vectorial problems of magnetohydrodynamics. A naive extension of these results suggests that people should gather in isotropic (circular) conglomerates. Indeed, cities normally have a roughly circular shape when there is no external restrictions such as those imposed by rivers or sea shores (anisotropy of environment). Of course, some cities can be extended for hundreds of kilometers along the sea shores or river banks (Sochi or Volgograd in the USSR) but the idea of a megapolis between Moscow and Leningrad (of length about 650 km) as proposed by the academician Dmitry Likhachev is generally considered a polemic exaggeration. This shape is typical of settlements extended along a road. A careful analysis of the evolutionary equation of bacterial evolution also reveals such extended structures at lower levels of intermittent hierarchy.

Linear structures can also arise in more delicate processes of the evolution of mankind. Thus, L.N. Gumilev has proposed a theory of the origin of ethnic communities which suggests that formation of ethnic groups occurs because of the concentration of a specific property, which Gumilev calls the passionarity, which, possibly, has a genetic nature.[e] Outbursts of ethnogenesis, which result in the formation of ethnic communities, are not randomly scattered in space and time. They occur simultaneously within narrow extended regions extending to distances comparable to the Earth's radius. Gumilev is inclined to propose an extraterrestrial origin of ethnogenetic outbursts. This conclusion seems to be premature; more likely, we find evidence of a linear, filamentary intermittent structure similar to that of the magnetic field.

A typical property of intermittent distributions is the violation of the ergodic properties: different types of averages cease to be coincident. In real life this violation is much more common than in the refined world of theoretical physics. Two examples of this violation come to mind. The genetic studies of the evolutionary process are based on space averages while predictions based on Mendel's law concern ensemble averages. For sufficiently large populations, these averages coincide (the Hardy-Weinberg law). However, in many isolated communities (e.g., in mountain valleys) plant populations are relatively small and their evolution deviates considerably from predictions of the Hardy-Weinberg law (see, e.g., Ville, 1967, (6359)). Another example (Luchinsky, 1984) considers economics. The average velocity of a vehicle is defined as the ratio of the distance travelled to the travel time,

$$\bar{v} = \frac{S}{T} = \frac{\int_0^T v(t)dt}{T} .$$

This velocity can naturally be considered as the time-average. However, one can consider the instantaneous velocity as a function of the coordinate measured along a trajectory, rather than a function of time,

[e] Ideas of Lev Gumilev (son of the famous poetess Anna Akhmatova and poet Nicolay Gumilev) are so radical that in the seventies his basic work *Ethnogenesis and the Biosphere of the Earth* could only be deposited as a manuscript (even though his shorter papers on this subject have been published). The publication of this monograph has now been announced.

which leads to the definition of spatially averaged velocity:

$$\langle v \rangle = \frac{\int_0^S v(s)ds}{S} \; .$$

It can be easily seen that when a vehicle moves with varying speed, $\langle v \rangle$ always exceeds $\overline{v}$. Notably, at Soviet motor transport depots the required amount of gasoline is determined based on one type of averaged velocity while the accounts of economic results are based on the other type. One can easily guess what happens with the resulting difference between the obtained and consumed amounts of gasoline.

Having mentioned the social applications of the theory of random processes, we cannot bypass the major catastrophes which recently stunned the world. One of them is the Challenger tragedy, the other is the Chernobyl accident. Here we discuss the Chernobyl catastrophe on which the authors are better informed. Another unfortunate event is the unexpected loss of control of the Phobos-1 space probe.

The idea of intermittency has induced a shift in our views on the role of exceptionally large fluctuations as introduced in physics by L. Boltzmann. Physics of the 20th century was rather sceptical about the importance of such rare events. However, modern scholars in other fields often admit very improbable events whose impact is very considerable and determines long-term evolution of the considered system. So, the historical development of some states, e.g., the Roman Empire, is sometimes regarded as an empire branched out from the general course of progress (see Stuerman and Trofimova, 1971). The geologists mention a new uniformism reviving the ideas of Cuvier. They consider that rare random events, like collision of the Earth with a large asteroid, have determined some crucial epochs in the evolution of life on the Earth (see Berggren and van Couvering, 1984). Gradually, one more possible catastrophe of this type, the nuclear holocaust, has become a subject of intense and considerable scientific discussion. Evidently, common concepts of probability based on everyday experience are of little help in such cases; human civilization can die in a nuclear war only once.

The expansion of the ideas of modern catastrophism causes a natural negative reaction of the scientific community, which is not used to such approaches. We often hear the question whether the implementation of the idea of a catastrophe into scientific usage means a rejection of scientific ways of cognition. It is an attempt to legalize the belief in miracles in science decorated by scientific phraseology. We believe that these ideas are principally different from such vulgar ideas as the explanation of Walking on the Sea by occasional directed coherent motion of the molecules which constitute the body. Indeed, there is an exceedingly small probability that in the course of their chaotic quasi-oscillatory motion, the molecules move altogether in one direction. We would not estimate this probability here; probably, it is much lower than all other small probabilities discussed in this book. We only note that even this improbable event can hardly lead to an expected result. Instead of walking, a heart attack due to a disturbed blood circulation is a more plausible result.

Improbable catastrophes which we discuss here admit two ways of interpretation: probabilistic and deterministic. The latter is more common and customary. When the chain of events that have led to a catastrophe is considered in detail, the disaster becomes a natural and unique consequence. To be specific, we briefly consider two recent improbable events of this kind, the Chernobyl accident and the loss of the Phobos-1 space probe. The reader can easily adapt our discussion to examples of which he (or she) is better informed, e.g., the Challenger accident.

The Chernobyl accident was an explosion which led to ejection of radioactive materials and contamination of vast areas, evacuation of about 100,000 inhabitants of the 30-km zone around the nuclear energy plant, health damage for thousands and mental shock for many more people living close to and far from Chernobyl – small town by Pripyat river, a tributary of Dnieper.

The immediate cause of the accident was an experiment planned beforehand during a scheduled shutdown of the reactor. In accordance with the plan of the experiment, one of the emergency control systems of the reactor was shut off although this violated the regulations governing maintenance of the reactor. However, this violation was necessary in order to carry out the planned experiment and was not dangerous by itself. Owing to a mistake made by an operator, the scheduled starting moment of the experiment was missed and the output power of the reactor had fallen too much. Even though the operator managed to turn on the output energy, one more emergency control system was switched off in order to avoid a complete shutdown of the reactor and to carry out the experiment. The computer of the control system then indicated that the reactor should be shut down immediately but, notwithstanding this, the operator had started the experiment. In fact, the conditions were far from those planned for this experiment. After a few seconds, the output power strongly increased and the heat transfer system became ineffective. An attempt to activate all available emergency control systems failed and the reactor was destroyed (see Information, 1986 for details).

Thus, the operator was determined to carry out the planned experiment under any circumstances. This motivated him to neglect all the regulations concerning normal handling of the reactor and led to the loss of the sense of danger. Paradoxically enough, such behaviour is known and understandable for a true experimentalist working hard in his laboratory. However, the Chernobyl reactor was not a scientific laboratory but only an industrial energy device which could not bear abnormal conditions. Probably, this contradiction between the limited possibilities of machinery and the freedom of a scientific experiment is a fundamental cause of the accident. In this sense, the accident is only a logical result of all the preceeding events and the role of chance reduces to that of an infernal driving belt.

However, all explanations of the particular causes of the Chernobyl accident do not explain that deep and strong reaction of the people worldwide which allows one to call this accident a catastrophe. This social aspect has been elucidated by the German philosopher Friedrich Schelling more than 150 years ago (see Schelling, 1972). As Schelling believed, the idea of permanent progress is essentially the idea of aimless progress and, therefore, infinite progress is a meaningless and sad concept. In accordance with this prophetic thought, the negative consequences of modern development of science and technology have brought mankind to the edge of disaster. The ghosts of nuclear holocaust and the ruined environment surround us. It is the responsibility of scientists to develop a new strategy of living which would rely on safe and ecologically pure solutions of current problems, thus diminishing the probability of such catastrophes. Nobody can suppress completely the role of randomness. We can only attempt to diminish the power of the chance.

Let us recall the other accident. In July 1988 two space probes, Phobos-1 and Phobos-2, were launched for Mars. The goal of the mission was to explore Mars and its satellite, Phobos. The space probe Phobos is a complicated autonomous system controlled by signals transmitted from the Earth. With an interval of approximately two days, a series of commands was transmitted to the probe, which were handled by onboard computers and executed by service systems and scientific instruments. The results of scientific measurements were transmitted back to Earth.

In early September an unexpected accident occured in the Phobos-1 spacecraft. Owing to a mistake, one character was omitted in one of the instructions transmitted to Phobos-1. Due to the occasional coincidence, the distorted command, which was originally devised for some scientific instrument, had caused the switching off of the orientation system of the solar-cell panels. As a result, the probe turned away from the Sun, the power supply went off and the probe ceased to obey commands from Earth. After a short period, an autonomous emergency storage battery had also discharged. For forty days, the personnel were unsuccessfully trying to re-activate the probe by uninterupted transmission of powerful radio signals. Phobos-1, which cost about 300 million dollars, was lost, together with the expected important scientific results. The fatal role was played by chance. Probably, the accident would not have occured if the commands were secure against distortion, texts of the commands were less similar, etc. However, nobody is safe from all possible sources of accidents, e.g., collision with a meteorite or accidents at the launch site. Space exploration will always remain a risky enterprise.

Both these accidents are (or will be) carefully analyzed by specialists. This will, in certain sense, reduce the probability of similar accidents. However, this approach, based on a traditional treatment of rare events, does not take an important peculiarity of improbable accidents into account: often they are not simply improbable and it is quite possible that a catastrophe of a given type would occur only once in the history of mankind. However, this does not imply that other catastrophes caused by effects still not envisaged by Man are impossible. Probably, this suggests that when planning a complicated modern technical system, the SDI complex or a chemical plant, we should be aware of the possibility of a catastrophic accident however reliable the safety measures seem to be. Of course, this does not imply that the safety measures are ineffective or that we should abandon the construction of the gigantic plants; the solution should be consciously motivated rather than dictated by illusions. As Epiktetos put it, we should take chances with bold confidence or otherwise stop the development. Both of these alternatives can be met in the history of mankind. The Titanic catastrophe has taught us that even the most reliable safety measures cannot ensure absolute safety of the vessels. Nevertheless, we supply every ship with life jackets and lifeboats in amounts sufficient for all the passengers and crew members and continue the navigation being conscious that these measures still cannot eliminate the risk but only decrease it. As an example of an alternative solution, we recall that in 1878, a leftist terrorist Vera Zasulich made an attempt on the life of the Petersburg city governor general Fyodor Trepoff. She was arrested and brough to a trial by jury. Her guilt was obvious but Zasulich was discharged by the jury whose members appreciated her motivation for the murder attempt. In those times this occasion was considered as a triumph of justice and until now remains an important argument in favour of court of jury. Meanwhile, the spread of the political air of terrorism in the 1970's has resulted in reconsideration of that early perception of terrorism: now we are inclined to condemn any terrorist act irrespective of its motivation.

The Chernobyl tragedy has induced a flood of papers, articles and books which discuss the details of the accident, guilt of the persons involved, their faults, etc. The technocrats and the antipollution activists argue about very specific problems. However, it seems that what is most important now is to develop a new thinking relevant to our nuclear age. From the standpoint of the concepts developed in this chapter, mininizing the consequences of such accidents requires an activation of nonlinear effects that would suppress intermittency. However, we would like to stress that the scale of this problem is so enormous that it can be solved only through combined efforts of politicians, artists, technocrats, and everybody else (see, e.g., Wolf, 1987).

CHAPTER

NINE

MAGNETIC FIELD IN A FLOW OF CONDUCTIVE FLUID

The previous chapter was devoted to the phenomenon of intermittency, i.e., to the development of intermittent structures in an initially smooth distribution. Intermittency is a very universal phenomenon which occurs practically irrespective of detailed properties of the background instability in a random medium provided only that the random field is of multiplicative type, because the basis of its development is provided by the central limit theorem. However, for a long period intermittency was neglected. The reason is associated, probably, with the fact that physics usually analyzes phenomena using simple and exactly solvable models. Such a model for studies of many instabilities is a deterministic medium with, e.g., spatially uniform parameters where intermittency cannot develop. In contrast to this general situation, the dynamo-generation of magnetic fields in a conducting medium cannot occur in simple and symmetric velocity fields. For this reason, almost from the very beginning of its development, the dynamo theory considers the random velocity field where intermittency is inevitable.

9.1. THE KINEMATIC DYNAMO PROBLEM

Since time immemorial, magnetism was a subject of great curiosity. In the Ancient World, the problem of explaining this phenomenon has got the reputation of being intractable. For instance, when discussing the problem of prediction and the study of random phenomena, Cicero mentioned in "De divinatione" the following example of an enigmatic but undoubtedly real phenomenon: "There exists a stone called magnet which attracts iron, but the cause of this I cannot explain; would you then deny the very fact?" (De div., I, 86). The explanation of this phenomenon of ferromagnetism is correctly considered as one of highest achievements of solid state physics. Ferromagnetism (as well as superconductivity) is an essentially quantum phenomenon and, therefore, exists only at low temperatures below the so-called Curie point. It is a fortunate coincidence that these low temperatures are sufficiently high from the human viewpoint (the Curie point of iron is about 1000 K). If high temperature superconductivity is a sensation of our days, high temperature ferromagnetism can be considered as a sensation of the era of Thales of Milet. Magnetism also has another *modus vivendi* as a component of an electromagnetic wave. This form of magnetism, which is so familiar in broadcasting, is essentially relativistic.

Magnetic fields also arise around wires that carry the electric currents. This last type of source of magnetic fields is frequently found in nature. Consider the Earth's magnetic field. We can be sure that the deep interior of the Earth does not hide a ferromagnetic crystal because the temperature there exceeds the Curie point. It is more realistic to envisage a ring current within the Earth, which produces the geomagnetic field. Magnetic fields exist also in many planets, the Sun, stars, galaxies and, probably, even in larger objects such as clusters of galaxies. In cosmic environments, quasi-stationary magnetic fields are one of the typical components. The range of the field strengths is very wide and, based on everyday experiences, in many cases one would call them weak. However, these magnetic fields are often in the state of what is called energy equipartition when the magnetic energy is comparable with the kinetic energy of small-scale motions. Therefore, the magnetic fields equally determine the dynamics of the cosmic matter.

In 1919, in his remarkable report to the British Association for Advancement of Science, J. Larmor formulated the problem of the explanation of cosmic magnetism. He correctly identified the source of the magnetic fields of the Earth and the Sun. Magnetic fields can be produced by the hydrodynamic motions of electrically neutral conducting mediums. As has been described by Faraday's induction law, the time variation of the magnetic field is determined by the curl of the electric field:

$$\frac{\partial \mathbf{H}}{\partial t} = -c\nabla \times \mathbf{E}\,.$$

As described by Ohm's law, in the reference system moving along with the medium, the electric current is proportional to the electric field. In the rest frame, the electric field has a corresponding magnetic field so that for the moving medium

$$\mathbf{E} = \sigma^{-1}\mathbf{j} - c^{-1}\mathbf{v} \times \mathbf{H},$$

where σ is the electric conductivity and $\mathbf{v}$ is the velocity of the medium. The velocity v is here assumed to be much smaller than the speed of light. Therefore, the displacement currents can be neglected and Oersted's law yields

$$\mathbf{j} = \frac{c}{4\pi} \nabla \times \mathbf{H}.$$

Elimination of j and H from these three equations yields the following equation which describes the generation of magnetic fields:

$$\frac{\partial \mathbf{H}}{\partial t} = \nabla \times (\mathbf{v} \times \mathbf{H}) - \nabla \times (\nu_m \nabla \times \mathbf{H}), \qquad (1)$$

where $\nu_m = c^2/(4\pi\sigma)$ is the magnetic diffusivity. This equation was first obtained by H. Hertz and it is known as the induction equation.

When considering the problem of the origin of magnetic fields, it is natural to assume that at the initial stages, magnetic fields are weak and do not affect the motions of the medium. Thus, the formulated problem in which the velocity field is prescribed is called the kinematic dynamo problem.

The maintenance of magnetic fields by electric currents flowing at infinity should be excluded. To do this, it is sufficient to require a sufficiently rapid decrease of the initial magnetic field $\mathbf{H}_0$ as $r \to \infty$.

The kinematic dynamo problem is linear in magnetic field strength. When we are interested in the self-excitation of the field (i.e., exponential time-growth), it is natural to compare the two terms on the right-hand side of equation (1). The second term describes magnetic dissipation while the first one can lead to field growth. For a crude estimation, we replace the derivatives by the corresponding ratios to obtain

$$\frac{\partial \mathbf{H}}{\partial t} \sim \frac{v\mathbf{H}}{l} - \nu_m \frac{\mathbf{H}}{l^2}, \tag{2}$$

where l is the characteristic scale of the velocity field. Thus, we see that field intensification prevails over dissipation when the magnetic Reynolds number

$$R_m = \frac{vl}{\nu_m}$$

is large. In cosmic plasmas, this condition is usually fulfilled; even though the cosmic plasma is not a very good conductor and plasma velocities are moderate, the involved distances are so large that the magnetic Reynolds number is usually very large, e.g., 10^8 in the Solar photosphere. Consequently, it can be expected that under such conditions the magnetic fields grow exponentially: $H \propto \exp(\gamma t)$ at the rate

$$\gamma \sim \frac{v}{l}, \tag{3}$$

which does not depend on the magnetic diffusivity for large magnetic Reynolds numbers. Such process of magnetic field growth is called fast dynamo. It would seem that only one more step is required to solve the problem: having considered some very simple flow, e.g., $\mathbf{v}(\mathbf{r}, t) = \text{const.}$, one should evaluate the numerical factor in (3) and the spatial distribution of the generated field. If it is successful, this solution would be a good basis for dealing with the more complicated velocity fields.

However, such a naive attempt cannot be successful. The simplest velocity field, parallel transformation, does not affect the magnetic field at all due to the Galilean invariance. Even more complicated, but still tractable, flows confined to planes or spherical surfaces are also a miss. Although for such flows, equation (1) cannot always be exactly solved, it can be shown that flows with a high degree of symmetry cannot generate magnetic fields (see, e.g., Moffatt, 1978).

Nevertheless, there exist steady velocity fields in which the embedded magnetic field grows exponentially. The simplest one is the helical flow of a conducting fluid in which the fluid both rotates around a certain axis and moves along it. This example can ensure that magnetic field can indeed be generated by a moving conducting fluid but the field growth rate of this dynamo cannot be estimated by (2) with the dissipation neglected. The reason for this is the fact that in this case the dissipation term does not only oppose field generation but also couples the different components of the magnetic field, which is necessary for its growth. Thus, magnetic diffusion controls the field growth, and for large R_m the growth rate turns out to be proportional to a certain negative power of the magnetic Reynolds number,

$$\gamma \sim \frac{v}{l} R_m^{-q}, \quad q > 0. \tag{4}$$

Such dynamos are said to be slow.

The physical nature of slow dynamos can be understood with the help of a procedure proposed by Alfven (1950) (see Fig. 9.1). A magnetic loop

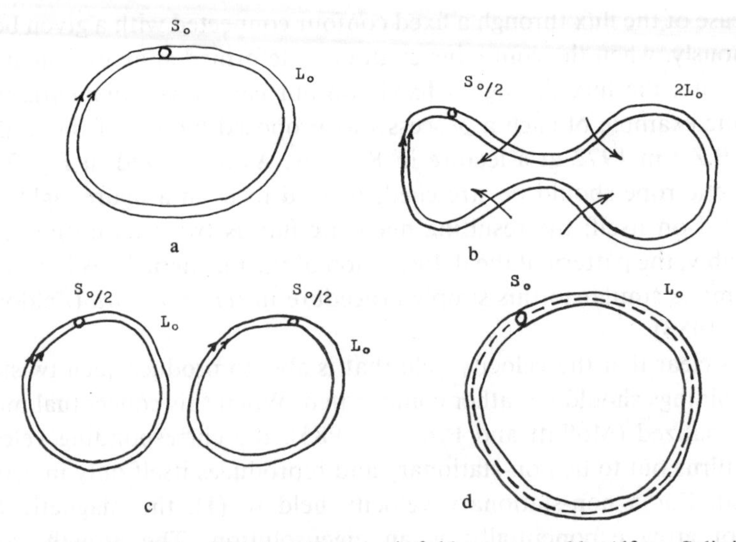

Fig. 9.1. Mechanism of amplification of magnetic field as proposed by Alfven. Splitting of the magnetic loop into two parts (the transition from (b) to (c)) occurs owing to magnetic diffusion.

is first stretched to twice its size. Then the segments with opposite directed fields are brought closer. Then the magnetic diffusion leads to separation of the loop into two loops whose sizes are equal to the initial loop size. Finally, these two loops are superimposed. As a result, the magnetic flux through the loop cross-section is doubled. Here we clearly see the role of magnetic diffusion: for $R_m \to \infty$ the field ceases to grow.

It cannot be excluded that such mechanisms of generation are of importance in some objects (e.g., jets ejected by radiogalaxies). But in the Sun, the magnetic fields vary at the time scale (3); this plausibly indicates that the fast dynamo is operative. Illustrative examples of the fast dynamo are considerably more complicated than Alfven's procedure. This can be seen from the fact that for an ideal conductor (infinite magnetic Reynolds number) the magnetic field is frozen into the medium and the magnetic flux through any fluid contour (comoving with the fluid) is conserved. It would seem that this implies that the number of magnetic lines that pass through a given contour can be doubled only with the help of diffusion, i.e., any dynamo is slow.

However, freezing of the field precludes a fast growth of the magnetic flux through a comoving contour while the dynamo can lead to an increase of the flux through a fixed contour connected with a given body. Obviously, when the comoving contour is deformed at an exponentially large rate, the flux through a fixed contour can grow exponentially. A specific example of such a process was proposed by one of the authors (Ya. B.Z.) in 1972 at a lecture in Krakow. As illustrated in Fig. 9.2, a magnetic rope should be stretched, twisted to form a figure eight, and doubled on itself; the resulting magnetic flux is twice its initial value. Probably, the pattern of the deformation of the magnetic lines in any fast dynamo is similar to this simple procedure in *rerum natura* (Zeldovich *et al.*, 1983).

It is clear that the velocity field that is able to produce such twistings and foldings should be rather complicated. When this conceptual model is formalized (Moffatt and Proctor, 1985), the corresponding velocity field turns out to be non-stationary and reproduces itself only in certain period. For a non-stationary velocity field in (1), the magnetic field cannot grow exponentially as an eigensolution. The growth rate γ characterizes, e.g., the time-averaged growth rate.

It is as yet unclear whether or not a stationary velocity field can act as a fast dynamo. The existing examples of a fast dynamo in a steady velocity

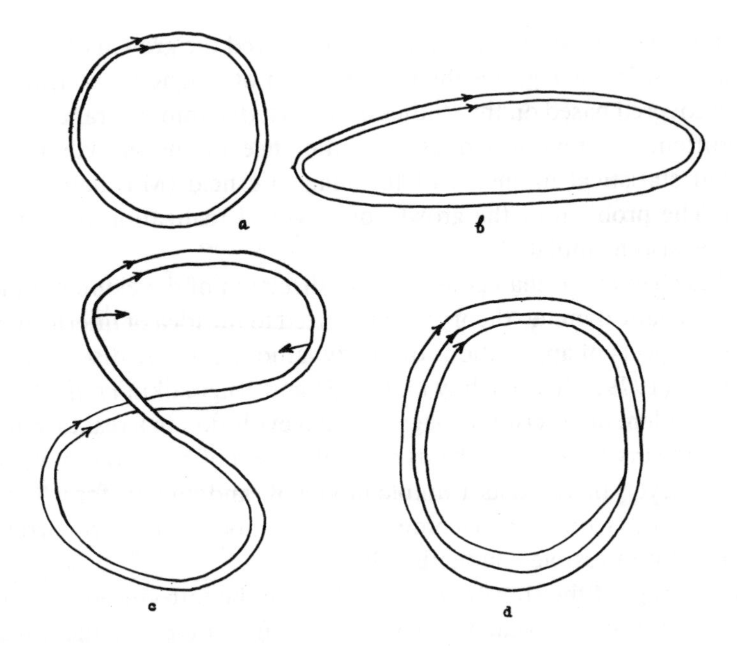

Fig. 9.2. The fast dynamo. The growth rate of the magnetic field here is independent of the magnetic diffusivity when the latter is small.

field are given either in a curved space (Arnold *et al.*, 1981) or in velocity fields with weakly singular vorticity (Soward, 1987). The only thing which is clear is the fact that these examples are too complicated to be a realistic flow model of cosmic plasmas, which are non-steady and, moreover, often turbulent.

9.2. THE SHORT-CORRELATED RANDOM FLOW

The situation with fast dynamo described in the last section tempts one to suggest a radical idea that the simplest flow that admits fast growth of the magnetic field is a random flow which is stationary only on average. This idea was first clearly proposed by M. Steenbeck, F. Krause and K.-H. Rädler (see Moffatt, 1978; Krause and Rädler, 1980) who considered the growth of magnetic field in a random flow as the growth of the mean magnetic field. This idea has allowed us to understand the origin of the large-scale magnetic fields in planets, stars and galaxies. Meanwhile, Batchelor (1950), Kazantsev (1967), Kraichnan and Nagarajan (1957),

Vainstein and Zeldovich (1972) have considered the growth of magnetic fluctuations. In the eighties the phenomenon of magnetic intermittency was discovered based on the studies of the fast dynamo in a random flow; intermittency is manifested as the difference of the growth rates of different statistical moments of the magnetic field (Molchanov *et al.*, 1985). The problem of the growth of a typical realization of magnetic field was also formulated.

In this chapter we shall consider that direction of development of the hydromagnetic dynamo theory which has led to the idea of intermittency. Various aspects of applications of the dynamo theory are discussed, e.g., by Parker (1979), Zeldovich *et al.* (1983) and Ruzmaikin *et al.* (1988).

The problem of describing magnetic field evolution in a real turbulence is by no means easier than the dynamo problem of a stationary velocity field. However, there exists a simple model of random flow for which the analysis of a dynamo can be completed. This model is a short-correlated flow of incompressible conducting fluid.

The concept of the short-correlated flow can be introduced as follows. Consider a random velocity field $\mathbf{v}_\Delta(\mathbf{r}, t, \omega)$ with ω as the random parameter, which is steady over every time interval $[n\Delta, (n+1)\Delta)$ of length Δ while velocity fields in any two different intervals are statistically independent. In other words, the flow *renovates* in every period Δ. To a certain extent, such flow can serve as a model of turbulent flows with exponentially decaying correlations. However, in contrast to turbulent flows, this flow is not exactly stationary on average since the period of variation of its statistical properties is equal to Δ. The short-correlated flow can be obtained as the limiting case of the renovating flow for $\Delta \to 0$. Of course, the velocity field of the short-correlated flow is a generalized function of time, analogous to Dirac's delta-function. When introducing the delta-function, one considers a sequence of functions $f_n(x)$ concentrated over a short-interval of length $\sim 1/n$ whose integrals are unity. The delta-function is understood as a limit of this sequence when the localization region shrinks to zero ($n \to \infty$). The integrals remain equal to unity only when $f_n \sim n$. Quite similarly, the limiting transition to the short-correlated flow means that the velocity v_Δ modulus is of the order $\Delta^{-1/2}$. This ensures a constant value of the mean squared velocity as Δ tends to zero.

Strictly speaking, the flows cannot be short-correlated in the same sense as there are no absolute black bodies or material points. A realistic criterion of closeness to the short-correlation model can be, e.g., the ratio of the renovation time Δ to the "eddy turnover" time l/v, where l and v are the characteristic scale and velocity of the flow, respectively. For a classical Kolmogorov turbulence, $\Delta \sim l/v$, and it cannot be approximated by a short-correlated flow. Nevertheless, the results of the short-correlated models can be reasonably extrapolated to the Kolmogorov turbulence because the correlation time Δ is usually several times smaller than the magnetic field growth time γ which, in turn, is somewhat larger than l/v.

It is remarkable, however, that the existent random flows sometimes deviate from the Kolmogorov type and the approximation of short correlations is quite reasonable for them. For instance, the turbulence in gaseous discs of galaxies is driven by the explosions of the supernova stars. The explosions are a result of physical processes inside stars and are practically independent of the interstellar processes. A resulting interstellar turbulent eddy evolves so slowly that before it decays another supernova explodes in its immediate vicinity, thus wiping out the memory of the previous explosion even before this occurs purely hydrodynamically. The ratio of the renovation time and the turnover time varies from one position to another in a galaxy. But in the central parts of galaxies, the short-correlation approximation seems to be adequate.

9.3. GENERATION OF THE MEAN MAGNETIC FIELD

The simplest result of the action of chaotic motions on a large-scale magnetic field embedded in a moving fluid is the tangling of magnetic lines. Due to the stretching of fluid elements, the magnetic energy can grow locally, but at first sight it can hardly be expected that the large-scale magnetic field would grow rather than decay. On the other hand, large-scale magnetic fields are a typical property of cosmic bodies. It turns out that the generation of large-scale magnetic fields by small-scale motions is associated with a special property of turbulence which is the mean helicity. In cosmic turbulent flows the numbers of vortices with left-handed and right-handed helicities are not equal. The average scalar product of the turbulent velocity and its curl, $\langle \mathbf{v} \cdot \nabla \times \mathbf{V} \rangle$, is called the mean helicity. The physical reason for non-vanishing of the mean helicity is the overall rotation of density-stratified cosmic bodies. When a turbulent vortex expands in a rotating volume, the Coriolis force produces an extra rotation so that vortices with some definite helicity become dominating. The mean helicity has opposite signs in the northern and southern hemispheres. Helicity is a pseudoscalar quantity; such quantity can be constructed from the angular velocity of a rotation $\mathbf{\Omega}$ and the density gradient: $\mathbf{\Omega} \cdot \nabla \rho$.

The mean helicity is a source of self-organization of the medium, it opposes the destructive action of turbulent tangling. Let us demonstrate this by taking a homogeneous, isotropic turbulence as an example. For the sake of simplicity here we consider the short-correlated velocity field. Rejection of this approximation leads to a considerably more complicated calculation but does not change the final conclusion that the large-scale magnetic field can be generated in helical turbulence.

The problem of evolution of the mean magnetic field in a random helical flow requires a derivation of the equation for the first statistical moment $\mathbf{B}(\mathbf{x}, t) = \langle \mathbf{H}(\mathbf{x}, t) \rangle$. For a short-correlated flow, this equation can be derived in the same manner as the equation for the moments of a scalar field in a short-correlated potential (Section 7.6). The resulting equation reads

$$\frac{\partial \mathbf{B}}{\partial t} = \nabla \times (\alpha \mathbf{B}) + \beta \Delta \mathbf{B}. \tag{5}$$

The coefficients α and β characterize the random flow. The former is proportional to the mean helicity, $\alpha = -\frac{\tau}{3} \langle \mathbf{v} \cdot \nabla \times \mathbf{v} \rangle$. The latter is the sum of the turbulent and molecular diffusions which play a leading role in transferring the admixtures in gases and fluids when directed flows are absent. It is the turbulent (or the closely associated convective) transfer which allows any scientist to feel the smell of a perfume after a few seconds when a lady enters his room. Taking for illustration the molecular diffusivity of ethyl spirit vapour in as air $\nu = 10^{-5}\ \mathrm{m^2 s^{-1}}$, we see that the spread of this admixture due to molecular diffusion in a room of the size $r = 3$ m takes a time

$$t = r^2/\nu \simeq 10^6 \text{ s} \simeq 10 \text{ days}.$$

Meanwhile, the air motions induced, e.g., by the lady walking along have a characteristic velocity 1 ms^{-1} and a scale 1 m, which implies that the turbulent diffusivity is 1 m^2s^{-1}. This diffusivity determines a spreading time of the order of 0.1 minute for the perfume in the room.

Let us verify that equation (5) includes the self-excitation of the mean magnetic field. For this purpose, consider the evolution of an initial magnetic field represented by a single Fourier mode:

$$\mathbf{b}(\mathbf{k}, t) = e^{\gamma(\mathbf{k})t}\mathbf{b}_0(\mathbf{k}), \quad \mathbf{B}(\mathbf{x}, t) = \mathbf{b}(\mathbf{k}, t)e^{i\mathbf{kx}}. \tag{6}$$

Substituting (6) into (5) we obtain the following equation for the growth rate $\gamma(k)$ and the generated field distribution:

$$\gamma(\mathbf{k})b_0(\mathbf{k}) = i\alpha\mathbf{k} \times b_0 - \beta k^2 b_0. \tag{7}$$

This system of three algebraic equations can be solved easily when one of the axes, e.g., the z-axis of the reference system, is chosen to be parallel to the wave vector $\mathbf{k}$. Then the solenoidality of magnetic field implies that $\mathbf{b}_0$ has only two non-vanishing components, b_x and b_y, so that (7) reduces to

$$\gamma b_x = -ik\alpha b_y - \beta k^2 b_x,$$

$$\gamma b_y = -ik\alpha b_x - \beta k^2 b_y.$$

This system of equations has solutions with a growth rate

$$\gamma = \alpha k - \beta k^2,$$

which is positive for $\alpha > 0$ and $k > \alpha/\beta$. The spatial field distribution can be described as a helical wave with right polarization. For negative α, the wave has left polarization and $\gamma = -\alpha k - \beta k^2$.

The helical magnetic waves with different wave vectors grow at different rates (Fig. 9.3). Self-excitation of the mean magnetic field in an infinite homogeneous medium is possible for any intensity of the helical motions: for arbitrarily small α, there exist long-wavelength modes for which the generating action of helicity dominates over the destructive action of diffusion. Therefore, the generated field is always the large-scale one. The field of the scale

$$k_0 = \frac{\alpha}{2\beta}$$

grows most rapidly. The corresponding growth rate is given by $\gamma(k_0) = \alpha^2/(4\beta)$. This means that an initial distribution of a weak magnetic field evolves into a distribution with distinguished scale $2\pi k_0^{-1}$, i.e., the field becomes structured.

In cosmic objects (stars, galaxies, etc.), magnetic field generation is more complicated. Apart from helicity, it depends on its differential rotation, the shape of the generation region, and other factors. However, many features of the simple picture described, particularly the magnetic structures, are preserved in these realistic dynamo systems (see Parker, 1979; Zeldovich *et al.*, 1983). For instance, various magnetic structures are observed in the nearby galaxies (Ruzmaikin *et al.*, 1988).

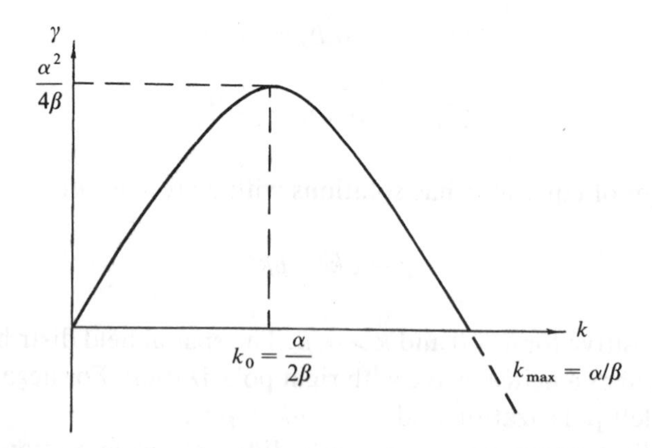

Fig. 9.3. The growth rate of the mean magnetic field versus the wave number k.

9.4. FLUCTUATIONS OF HELICITY

The problem of self-excitation of magnetic field by the motions of turbulent fluid was motivated by the inquiry into the origin of magnetism of planets, stars and galaxies. In these applications, most interesting are the large-scale magnetic fields which can be distinguished against the background of the fluctuation fields generated simultaneously.

Also of interest is the question of magnetic field generation by bulk motions of liquid metal in industrial devices, e.g., breeder reactors (Kirko *et al.*, 1981; Frisch *et al.*, 1985) and metallurgical installations (Mestel, 1984). These flows are characterized by the Reynolds numbers of the order of 10^7 and, therefore, they are turbulent. The magnetic Reynolds numbers are not so large and do not exceed 10–100; nevertheless, these values are sufficient to make possible the growth of magnetic fields. Unlike astrophysical conditions, here the relevant volumes are only moderate and the overall rotation is absent; both these factors play an outstanding role in the generation of astrophysical large-scale magnetic fields. In the abovementioned industrial devices, excitation of small-scale, fluctuational magnetic fields is plausible. This conjecture is substantiated by experimental results (Kirko *et al.*, 1981) which indeed suggest generation of magnetic fields, but of another type than that predicted by the mean-field dynamo theory based on the mean helicity of turbulence.

The problem of self-excitation of magnetic field in a flow which is reflectionally symmetric on average is also important for some astrophysical objects, e.g., the intergalactic gas in galaxy clusters.

This problem has proved to be considerably more complicated than the problem of the origin of the mean magnetic field in a helical random flow. Although the pioneering paper of Batchelor (1950), devoted to self-excitation of fluctuative magnetic fields in a non-helical turbulent flow, preceded by several years the papers of Parker and Steenbeck *et al.*, which developed the idea of helical turbulent mean field dynamo, systematic study of the former problem has not been completed up to now. Partly, the complexity of this problem is associated with the fact that the following two approaches, which seem to be equivalent, turn out to be basically different. The first approach in studying the small-scale magnetic fields consists in the derivation and analysis of the equation governing their correlation tensor. Such an equation was first derived by Kazantsev (1967) and Kraichnan and Nagarajan (1967) in the case of the short-correlated flow. The results of such analysis are discussed in Section 9.6 below.

Another seemingly equivalent approach is due to Kraichnan (1976), who has proposed the following tempting idea as to how to bypass the complexities of analysis of the equation for the correlation tensor. The vanishing of the mean helicity does not imply that $\mathbf{v} \cdot \nabla \times \mathbf{v}$ identically vanishes everywhere. Such vanishing occurs only for relatively simple degenerated flows. In general, vanishing of the mean helicity occurs owing to compensations of the values of helicity which acquire different signs and values for different moments, positions and different realizations of a random flow. This observation suggests the idea of considering the generation of magnetic fields by fluctuations of the helicity. More rigorously, one may consider a turbulent flow characterized by two random parameters (e.g., characterizing the turbulent motion within a star and stellar encounters in the example below). Even though this problem is not motivated by any particular experimental or observational results, it seems reasonable in connection with, e.g., the random flows within a rotating star. Such flow has a mean helicity whose sign is determined by the direction of the star's rotation. Consider a star that is subject to external forces and therefore changes the modulus and direction of its angular velocity (for instance, this can be a star near the center of a globular cluster where gravitational encounters are frequent). If the relaxation time of the internal random flows is shorter than the time between the encounters, the helicity averaged over both the random factors (the intrinsic hydrodynamic random ensemble and the external force) is zero. Let us consider the evolution of the magnetic field in such a random medium.

Partial averaging of the magnetic field over the ensemble of realizations of random hydrodynamic flows within the star leads to equation (5). The next step is averaging over the external forces.

For the sake of simplicity, consider a uniform and constant turbulent diffusivity β. We also assume that α is uniform and renovates with period τ, and $\langle \alpha \rangle = 0$.

The magnetic field in a random medium with two independent random parameters can be averaged over either of them separately. We denote the field averaged over the turbulent ensemble within a star by **B**, while the result of the additional averaging over helicity fluctuations is denoted by $\langle \mathbf{B} \rangle$. Consider an initial field in the form of a helical wave with the wave vector directed along the z-axis and positive helicity: $\nabla \times \mathbf{B}_0 = +k\mathbf{B}_0$, $\mathbf{B}_0(z) = (\sin kz; \cos kz; 0)$. Substituting in (5) $\mathbf{B}(\mathbf{x}, t) = b(t)\mathbf{B}_0(\mathbf{x})$, we obtain the following equation for $b(t)$:

$$\frac{\partial b}{\partial t} = (\alpha k - \beta k^2) b,$$

where $\alpha(t, \omega)$ is the random coefficient. Evidently, its solution is given by

$$b(t) = b_0 \exp\left(-\beta k^2 t + k \int_0^t \alpha(s, w)\, ds \right). \tag{8}$$

Following the central limit theorem for $t \to \infty$, the integral in (8) can be estimated as $\int_0^t \alpha\, ds \simeq \langle \alpha \rangle\, t + \tau^{1/2} \langle \alpha^2 \rangle^{1/2} \xi t^{1/2}$, where ξ is a random quantity with unit dispersion. Since $\langle \alpha \rangle = 0$, this integral grows as $t^{1/2}$. Consequently, the solution $b(t)$ decays asymptotically. On the other hand, the mean field grows for $\langle \alpha^2 \rangle^{1/2} > 2\beta$. This can be seen from the fact that for small k we have approximately

$$\langle b(t) \rangle = b_0 \exp\left(-\beta k^2 t + \frac{k^2 \langle \alpha^2 \rangle \tau t}{2} \right), \quad t \to \infty \tag{9}$$

where we have used the fact that ξ is a Gaussian random quantity with zero mean value and unit dispersion. Relation (9) is exact for a short-correlated helicity.

Thus, fluctuations of α introduce a negative contribution to the diffusivity of magnetic fields. This becomes especially obvious when, using (9), the equation for the mean magnetic field is recast as

$$\frac{\partial \langle \mathbf{b} \rangle}{\partial t} = \left(\beta - \frac{\langle \alpha^2 \rangle \tau}{2} \right) \Delta \langle \mathbf{b} \rangle. \tag{10}$$

One can easily verify that the higher moments of $\langle b^p \rangle$ grow progressively, i.e., the field distribution is intermittent.

These results are valid for the particular form of initial field. In fact, an arbitrary initial field can be expanded in helical waves with various directions of the wave vector k and signs of the waves' helicity. Notice that for the negative-helicity waves of the type $\mathbf{B}_0(\mathbf{x}) = (\cos kz, \sin kz, 0)$, the sign of the second term in the exponent of (8) changes. However, this does not change the behaviour of the mean field. In a generic situation, due to the contribution of modes with small k, a typical realization of the field decays as a power law (as $t^{-3/2}$ in three dimensions) rather than exponentially. The statistical moments also show progressive growth.

Such behaviour occurs for moderate values of helicity dispersion, $\langle \alpha^2 \rangle \leqslant \langle \alpha^2 \rangle_{cr} = 2\beta/\tau$. When $\langle \alpha^2 \rangle$ exceeds this critical value, the diffusion coefficient in (10) becomes negative. Taken literally, this means that the super-exponential growth of the mean field is accompanied by a large decrease in its scale. This process is exactly the inverse of the well known diffusive spreading. However, the short-correlation approximation soon becomes inapplicable in this case and field evolution depends on the detailed properties of the helicity correlation function.

9.5. HIGHER STATISTICAL MOMENTS AND INTERMITTENCY IN THE FAST DYNAMO

Let us return to random flows characterized by a single random parameter. As we have already noted, the study of the properties of self-excited magnetic field assumes the derivation of the moment equations for the magnetic field and the investigation of their properties. These equations usually are very cumbersome and it is instructive to begin with the case of small magnetic diffusivity where phenomena of the type of intermittency are more strongly pronounced (Molchanov *et al.*, Zeldovich *et al.*, 1987).

Let us explicitly evaluate the growth rates of a typical realization of a magnetic field and all its moments in the limiting case of large magnetic Reynolds numbers and flows with short, δ-function-like time correlations. The existing theorem for solving the induction equation and some estimates of the growth rates for such a flow have been obtained by Rozovsky *et al.* (1984).

The adopted asymptotic method is based on the fact that the growth rates of magnetic field and its moments in three-dimensional random flows, in the limit of a large magnetic Reynolds number, coincide with the growth rates calculated in neglect of magnetic diffusion (Molchanov *et al.*, 1984). This makes the Lagrangian solution of the induction equation useful. Thus, the problem reduces to the analysis of the field evolution $\mathbf{H}(\mathbf{x}, t)$ along the Lagrangian path

$$\xi_t = \mathbf{x} + \int_0^t \mathbf{v}(\xi_s, s)\, ds,$$

determined solely by the prescribed divergence-free velocity field $\mathbf{v}(\mathbf{x}, t)$. According to the induction law, in a short period Δt the field changes according to

$$H_i(\xi_{t+\Delta t}, t + \Delta t) = \left(\delta_{ij} + \frac{\partial v_i(\xi_t, t)}{\partial x_j}\Delta t\right) H_j(\xi_t, t), \quad i, j = 1, 2, 3. \quad (11)$$

Consider the velocity field with a vanishing average velocity. Then the property of δ-correlation of the velocity field is equivalent to every element of the matrix $W_{ij} \equiv (\partial v_i/\partial x_j)$ being a Wiener random process with zero mean value and a certain dispersion. Additionally we suppose that the velocity field is statistically homogeneous and has an isotropic probability distribution. Then the matrix W_{ij} can be represented as

$$W_{ij} = aw\delta_{ij} + \sigma w_{ij}, \quad (12)$$

where a and σ are constants while w and w_{ij} are independent Wiener processes with unit dispersion. The first term on the right-hand side of (12) is obviously isotropic. Generally speaking, any particular realization of the second matrix is anisotropic. However, the distribution of w_{ij} is isotropic. To show this, we prove that this tensor distribution remains unaffected under orthogonal transformations. Since the matrix w_{ij} is Gaussian, it is sufficient to transform it by the orthogonal matrix U_{il}, $\tilde{w}_{ij} = U_{ik} U_{lj}, W_{kl}$, and verify that the correlator $\langle \tilde{w}_{ij} \tilde{w}_{mn} \rangle$ coincides with

$\langle w_{ij} w_{mn} \rangle$. The solenoidality of the velocity field additionally implies that the trace of matrix W_{ij} vanishes. This gives

$$W_{ij} = \sigma \left(w_{ij} - \frac{1}{3} w_{ll} \delta_{ij} \right),$$

$$\overline{\frac{1}{2} \frac{\partial v_i}{\partial x_j} \frac{\partial v_i}{\partial x_j}} (\Delta t)^2 = \langle W_{ij} W_{ij} \rangle \;\; = 8\sigma^2 \Delta t, \tag{13}$$

$$\overline{\frac{1}{2} \frac{\partial v_i}{\partial x_j} \frac{\partial v_k}{\partial x_l}} (\Delta t)^2 = \langle W_{ij} W_{kl} \rangle \;\; = \sigma^2 \left(\delta_{ik} \delta_{jl} - \frac{1}{3} \delta_{ij} \delta_{kl} \right) \Delta t,$$

where w_{ll} is the trace of matrix w_{ij}. The symbol $\langle \ldots \rangle$ here denotes averaging over the Wiener process and the bar denotes averaging over the velocity field distribution.

It would seem that the expression (12) should be supplemented by one more term of the form $\varepsilon_{ijk} w_k$, with ε_{ijk} being the unit antisymmetric tensor. Since the Wiener process is invariant under reflection of spatial coordinates, this term has an isotropic probability distribution but, in contrast to w_{ij}, it is a pseudo-tensor rather than a tensor. Note that the helicity of a flow, which is an essential property for magnetic field generation, is connected with the pseudo-scalar part of the velocity correlation tensor. In terms of the matrix W_{ij}, the helicity is expressed through correlators of the type $\langle W_{ij} v_k \rangle$.

Let us derive the equation that describes the evolution of the squared magnetic field along a Lagrangian path. From (12)–(13) we obtain

$$H^2(t + \Delta t) = H^2(t) + \left(2W_{ij} \frac{H_i H_j}{H^2} + W_{ij} W_{ik} \frac{H_j H_k}{H^2} \right) H.$$

The quantity $2W_{ij} H_i H_j / H^2$ is a linear combination of Wiener processes and therefore this is also a Wiener process with zero average value and a dispersion $8\sigma^2/3$, which follows from (13). Since the correlation time of the Wiener process is short compared with Δt, the central limit theorem allows us to replace the second term in the parentheses by $8\sigma^2\Delta t/3$. Finally, the limit $\Delta t \to 0$ yields the desired equation:

$$\frac{dH^2}{H^2} = \left(\frac{8}{3}\right)^{1/2} \sigma w dt + \frac{8}{3}\sigma^2 dt.$$

The solution to this equation is given by

$$H^2(t) = H^2(0) \exp\left[\left(\frac{8}{3}\right)^{1/2} \sigma w_t + \frac{8}{3}\sigma^2 t\right]. \tag{14}$$

Since w_t grows only as $t^{1/2}$, for $t \gg \sigma^{-2}$ the first term in the exponent may be neglected. Thus, the growth rate of the magnetic field is

$$\gamma \equiv \lim_{t\to\infty} \frac{\ln H}{t} = \frac{4}{3}\sigma^2.$$

Now we can evaluate the growth rate of the $2p$-th moment of the magnetic field modulus. Raising expression (14) to the p-th power and averaging over the velocity field with the use of the relation $\langle \exp(pw_t) = \exp(p^2 t/2) \rangle$, we obtain

$$\gamma_{2p} \equiv \lim_{t\to\infty} \frac{\ln \langle H^{2p} \rangle}{t} = \frac{8}{3} p\sigma^2 + \frac{4}{3} p^2\sigma^2, \tag{15}$$

$$\frac{\gamma_{2p}}{2p} \equiv \frac{4\sigma^2}{3} + \frac{2\sigma^2 p}{3}$$

for arbitrary p. For a homogeneous isotropic flow with longitudinal correlation function

$$F(r) = \exp\left[-\frac{3}{5}\left(\frac{r}{c}\right)^2\right],$$

we have

$$\overline{\frac{\partial v_i}{\partial x_j}\frac{\partial v_i}{\partial x_j}}(\Delta t)^2 = 6\frac{v}{l}\Delta t,$$

where v and l are the characteristic velocity and scale of the flow. Thus, $\gamma_2/2 = 3/4\, v/l$ for the flow.

We have derived the rates of growth of the field and its moments along the Lagrangian trajectory. Due to homogeneity, the growth rates at a given Eulerian position $\mathbf{x}$ are the same.

Now we can easily understand how the magnetic field evolves in a short-correlated and unsteady average medium. In this case σ becomes a function of time so that $\sigma^2 t$ in (14) should be substituted by its average value over the considered period:

$$\int_0^t \sigma^2(s)\, ds\,.$$

Thus, when σ decreases with t rapidly enough, the field growth can become a power law in time and the intermittent strong gradients of the field strength cease to grow.

For $\nu_m = 0$, the spatial structure of the moments is described by the eigenvectors of a certain operator A_p derived by Molchanov *et al.*, (1987). The corresponding random magnetic field is a generalized function of position. In order to obtain the structure of the moments for small but finite ν_m, the next approximation in ν_m should be invoked. This kind of derivation of the growth rates and the corresponding eigenfunctions in two successive orders of the perturbation theory is typical, e.g., of the quasi-classical approximation (Maslov and Fedorjuk, 1981).

9.6. STRUCTURE OF THE MAGNETIC FIELD

We have seen that for a short-correlated flow of conducting fluid with a high magnetic Reynolds number, the self-excited field is such that the second, the fourth and higher moments of the magnetic field are determined not by the typical values of the growing field but rather by its rate realizations (outbursts). In order to elucidate the structure of the field, one should consider the moment equations in more detail. Here we consider the second moment.

In a inhomogeneous, anisotropic short-correlated flow, this equation can be obtained using the technique of Wiener integration and expansion in short correlation time, as described by Molchanov *et al.* (1985). This somewhat cumbersome equation has the following form:

$$\frac{\partial H_{ij}(\mathbf{x}, \mathbf{y}, t)}{\partial t} = \left\langle \frac{\partial v_i(\mathbf{x})}{\partial x_l} \frac{\partial v_j(\mathbf{y})}{\partial y_m} \right\rangle H_{lm} + \frac{1}{2} \frac{\partial}{\partial x_k} \left\langle v_i \frac{\partial v_k}{\partial x_l} - v_k \frac{\partial v_i}{\partial x_l} \right\rangle H_{lj}$$
$$+ \frac{1}{2} \frac{\partial}{\partial y_k} \left\langle v_i \frac{\partial v_k}{\partial y_l} - v_k \frac{\partial v_i}{\partial y_l} \right\rangle H_{il} - \left\langle v_k \frac{\partial v_i}{\partial x_l} \right\rangle \frac{\partial H_{ij}}{\partial x_k}$$
$$- \left\langle v_k \frac{\partial v_j}{\partial x_l} \right\rangle \frac{\partial H_{il}}{\partial y_n} - \left\langle v_k(\mathbf{y}) \frac{\partial^2 v_i(\mathbf{x})}{\partial x_l} \right\rangle \frac{\partial H_{lj}}{\partial y_k}$$
$$- \left\langle v_k(\mathbf{x}) \frac{\partial v_j(\mathbf{y})}{\partial y_l} \right\rangle \frac{\partial H_{il}}{\partial x_k} + \langle v_l(\mathbf{x}) v_m(\mathbf{y}) \rangle \frac{\partial^2 H_{ij}}{\partial x_l \partial x_m}$$
$$+ \left(\nu_m \delta_{kl} + \frac{1}{2} \langle v_k v_l \rangle_{\mathbf{x}} \right) \frac{\partial^2 H_{ij}}{\partial x_k \partial x_l}$$
$$+ \left(\nu_m \delta_{kl} + \frac{1}{2} \langle v_k v_l \rangle_{y} \right) \frac{\partial^2 H_{ij}}{\partial y_k \partial y_l}, \tag{16}$$

where

$$H_{ij}(\mathbf{x}, \mathbf{y}, t) \equiv \langle H_i(\mathbf{x}, t) H_j(\mathbf{y}, t) \rangle$$

(Molchanov *et al.*, 1983). Let us consider this equation in the case of statistically homogeneous, isotropic and reflectionally invariant velocity field whose correlation tensor is given by

$$\langle v_i(\mathbf{x}) v_j(\mathbf{y}) \rangle = \frac{l}{v} V_{ij}(\mathbf{x}, \mathbf{y}) = \frac{lv}{3} \left[F(r)\delta_{ij} + \frac{r}{2}\frac{dF}{dr}\left(\delta_{ij} - \frac{r_i r_j}{r^2}\right) \right], \quad (17)$$

$$\mathbf{r} = \mathbf{x} - \mathbf{y}, \qquad r = |\mathbf{r}|,$$

(Monin and Yaglom, 1971) where we have singled out the dimensional factors $l = \int_0^\infty F(r)dr$, the correlation length and the r.m.s. velocity fluctuation. The longitudinal correlation function is dimensionless, it equals unity at the origin, has a positive Fourier transform and decays at infinity.

Let us consider the solution that grows at the greatest rate. It is natural to presume that it has the same symmetries as the velocity field:

$$H_{ij}(r, t) = \frac{1}{3} W(r, t)\, \delta_{ij} + \frac{r}{2}\frac{dW}{dr}\left(\delta_{ij} - \frac{r_i r_j}{r^2}\right). \quad (18)$$

As shown by Kazantsev (1967), this assumption reduces (16) to an equation which is close to the Schrödinger equation but features a variable mass and has no imaginary unity in the time-derivative term.

This equation takes a simpler form when it is formulated in terms of an auxiliary function $\psi(r, t)$. This is related to the longitudinal correlation function of magnetic field W through the relation

$$\psi(r, t) = \frac{r^2}{3(2m(r))^{1/2}} W(r, t)\,, \quad (19)$$

where r is the separation of the positions for which the correlation is evaluated and $m(r)$ is the variable "mass" defined through

$$\frac{1}{2m} = \varepsilon + \frac{1 - F(r)}{3}.$$

Here ε is the inverse magnetic Reynolds number. Below, all quantities are made dimensionless by being measured in l, v and their combinations. Kazantsev's equation reads

$$\frac{1}{2}\frac{\partial\psi}{\partial t} = \frac{1}{2m(r)}\frac{\partial^2\psi}{\partial r^2} - U(r)\psi, \qquad (20)$$

where the potential $U(r)$ is given by

$$U(r) = \frac{1}{2r} - \frac{df}{dr} + \frac{1}{mr^2} - \frac{1}{8m^3}\left(\frac{dm}{dr}\right)^2,$$

with

$$f(r) = \frac{1}{3r^2}\frac{d}{dr}(r^3F).$$

The qualitative dependencies of the effective mass m and potential U are shown in Fig. 9.4.

The solution to equation (20) is sought in the form $\psi(r, t) = \psi(r)\exp(2\gamma t)$ which is equivalent to the formulation of the boundary-value

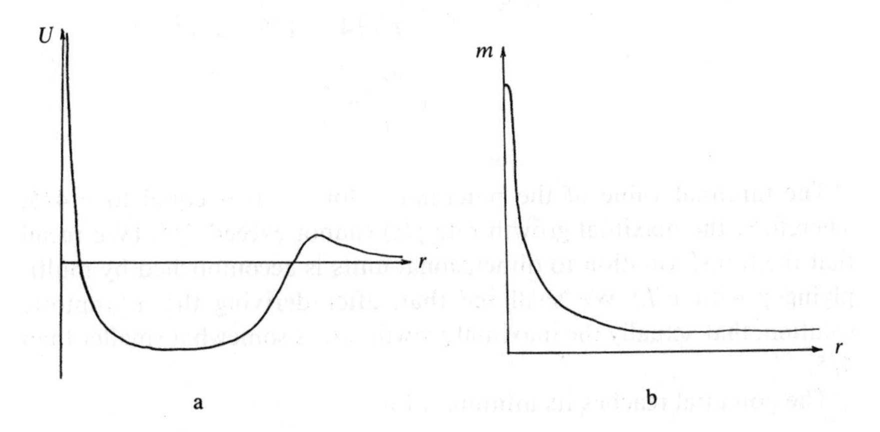

Fig. 9.4. The forms of potential (a) and mass (b) in the problem of self-excitation of the second moment of magnetic field.

problem for eigenfunction $\psi(r)$ in potential $U(r)$ with boundary conditions $\psi(0) = \psi(\infty) = 0$. To be more exact the condition $W(0) = 1$ reduces to $\psi \sim r^2$ for $r \to 0$ (see (19)). The existence of growing solutions of the equation (20) was shown by Kazantsev (1967) and Vainstein (1980). A detailed study of the problem was accomplished by Novikov *et al.* (1983) and Maslova *et al.* (1985) for a number of typical forms of the function $F(r)$. Here we present the asymptotic solution of equation (20) for the considered limiting case $\varepsilon \to 0$ in the special case $f(r) = \exp(-r^2)$ (Artamonova and Sokoloff, 1986). The following properties of $f(r)$ are essential: normalization conditions $f(0) = 1$ and $f''(0) = -2$, and smoothness $f'''(0) = 0$. It is natural to expect that in the limit $\varepsilon \to 0$ the solution concentrates at small r, i.e., near the bottom of the potential well $U(r)$, so that $U(r)$ and $(2m)^{-1}$ can be replaced by

$$
\begin{aligned}
f &\simeq 1 - r^2 + \frac{r^4}{2}, \\
\frac{1}{2m} &\simeq \varepsilon + \frac{r^2}{5} - \frac{r^4}{14}, \qquad (21) \\
U &\simeq -\frac{3}{5} + \frac{1/25 + 2\varepsilon/7}{r^2/14 - 1/5 - \varepsilon/r^2} \\
&\quad + \frac{8r^2}{7} + \frac{2\varepsilon}{r^2}.
\end{aligned}
$$

The minimal value of the potential U for $\varepsilon = 0$ is equal to $-4/5$. Therefore, the maximal growth rate $\gamma(\varepsilon)$ cannot exceed $4/5$. (We recall that the transformation to dimensional units is accomplished by multiplying γ with v/l.) We shall see that, after deriving the asymptotic solution, that actually the maximal growth rate is somewhat smaller than $4/5$.

The potential reaches its minimum for

$$r_{\min} = \left(\frac{14}{5}\varepsilon\right)^{1/4}$$

and the corresponding potential well is rather wide. Since we expect the solutions to concentrate near this point, it is natural to introduce a new stretched variable $\xi = r\varepsilon^{-1/4}$ so that in the new variable the potential is a slowly varying function of ξ even at the boundary of the well. Note that the introduction of the scale $\varepsilon^{1/4}$ is rather unusual and farther from the flat bottom of the well appears another natural scale $\varepsilon^{1/2}$ corresponding to a boundary layer at $r = 0$.

In terms of the new variable,

$$U(\xi) = -\frac{4}{5} + \varepsilon^{1/2} u(\xi),$$

where

$$u(\xi) = \frac{3}{\xi^2} + \frac{15}{14}\xi^2.$$

We see that the potential is nearly constant near the minimum point where the eigenfunction is concentrated. Its variation is pronounced either at small r ($\xi \sim \varepsilon^{1/4}$, i.e., $r \sim \varepsilon^{1/2}$) or for $r = 0(1)$ ($\xi \sim \varepsilon^{-1/4}$).

After omitting a small variable part of the potential we obtain the following equation

$$\xi^2\psi'' + (4 - 5\gamma)\psi = 0, \tag{22}$$

which can be easily solved.

Now the solutions to equation (22) should be matched with the solutions at small r of the order $\varepsilon^{1/2}$ (inner region) and large r of the order of unity (outer region), i.e., with the solutions in the regions where variations of the potential are considerable. The solutions for the inner and outer regions are found to be

$$\psi(r) = \begin{cases} B(r\varepsilon^{-1/2})^2, & r \leqslant r_1 = a\,\varepsilon^{1/2}, \\ C\exp\left(-\frac{3}{2}r\right), & r \geqslant r_2 = b, \end{cases}$$

where a and b are certain constants whose values are determined below. As usual, both the function ψ and its derivatives should be matched. The solution of equation (22) is such that the derivatives of ψ at $r = r_1$ and $r = r_2$ are smaller by a factor $\ln \varepsilon$ than ψ itself. Consequently, the matching should be accomplished simultaneously in two orders of asymptotic expansion in ε. To the first order, we seek the function $\psi^{(1)}(r)$ which satisfies the equation (22), vanishes at r_1 and r_2 and is identically zero in the inner and outer regions. Afterwards, we compensate for the mismatch of the derivatives of $\psi^{(1)}$ at r_1 and r_2 with the use of the second-order correction $\psi^{(2)}$ at $r_1 \leqslant r \leqslant r_2$ and the leading-order solutions in the inner and outer regions; simultaneously, matching of the second-order approximation of the eigenfunction is accomplished. As a result, the asymptotic eigenvalue is obtained from the leading-order equations while the eigenfunction which is satisfactory in all three regions is obtained from the equations of two lowest orders.

We note that this asymptotic procedure is a variant of the quasi-classical approximation for which the derivation of the eigenvalues and the corresponding eigenfunctions from the equations of two consequent orders is rather typical (Maslov and Fedorjuk, 1981). The constants a and b do not enter the first-order solution because the term $\ln(a/b)$, to which they combine, is much smaller than $\ln \varepsilon$; therefore, we can put $a = b = 1$ without loss of generality.

The procedure described above leads to the following leading-order solution:

$$\gamma_n = \frac{3}{4} - \frac{1}{5}\left(\frac{2\pi n}{\ln \varepsilon}\right)^2,$$
$$\psi_n = A_n r^{1/2} \sin\left(\frac{2\pi n}{\ln \varepsilon} \ln r\right), \tag{23}$$

where $n = 1, 2, 3, \ldots$ enumerates the eigenfunctions (cf. (15)).

Now we are in a position to evaluate the growth rate of the second moment of the magnetic field for vanishing magnetic diffusivity. Let us put $\varepsilon = 0$ in (21) and determine the upper bound of the spectrum for the corresponding equation (20). We expect again that the solution is concentrated at small r. We therefore introduce a new variable $x = r/a$

with $a \ll 1$. To the leading order in a^{-1} we again obtain equation (22) which does not include a. Introducing the other variables $\psi = x^{1/2}p(x)$ and $x = \exp y$, we obtain for $\psi(y)$ the one-dimensional Laplace equation

$$\frac{d^2\psi}{dy^2} = \left(5\gamma(0) - \frac{15}{4}\right)\psi,$$

whose spectrum has zero upper bound. Consequently, $\gamma(0) = 3/4$. Comparing this result with (23), we see that $\gamma(\varepsilon)$ continuously matches $\gamma(0)$ for $\varepsilon \to 0$ even though γ approaches this limit only logarithmically.

The limiting transition $\gamma(\varepsilon) \to \gamma(0)$ occurs only for a three-dimensional velocity field. For two-dimensional velocity fields, the potential $U(r)$ is positive everywhere (see Appendix in Novikov *et al.*, 1983). Thus, in two dimensions for $\varepsilon \neq 0$, the second moment of magnetic field experiences only a temporal growth (Zeldovich, 1956) with subsequent exponential decay.

Matching the asymptotic solutions to the second order, one can complete the derivation of the eigenfunctions of Kazantsev's equation and elaborate the expression for the eigenvalue by adding the next approximation γ_2. We do not give here the corresponding bulky equations (see Kleeorin *et al.*, 1986) but only illustrate them graphically.

The correlation function which corresponds to the leading mode is shown schematically in Fig. 9.5. The scale $r_1 \sim R_m^{-1/2} l$ corresponds to the skin-layer thickness with $W(r_1) = 0.82$. At $r = r_2$ the correlation function vanishes while at $r = r_3$ it attains its minimal value. The exact values of r_2 and r_3 depend on the detailed structure of $f(r)$ as well as the exact distance $r_4 = l$ at which the correlation function becomes exponentially decaying and $W(r_4) \approx -R_m^{-5/4}$ (recall that $R_m \equiv \varepsilon^{-1}$).

We propose the following interpretation of these characteristic scales based on the presumption that the stretch-twist-fold rope dynamo of Fig. 9.5 represents the basic process of field amplification in turbulence. The range (r_2, r_4) corresponds to the scatter in lengths of the initial loop of the magnetic lines. The mode number corresponds to the number of full rotations which the loop experiences when being twisted. The distance r_2 determines the size of the folded part of the loop. The anti-correlated tail is due to the solenoidality condition of magnetic field

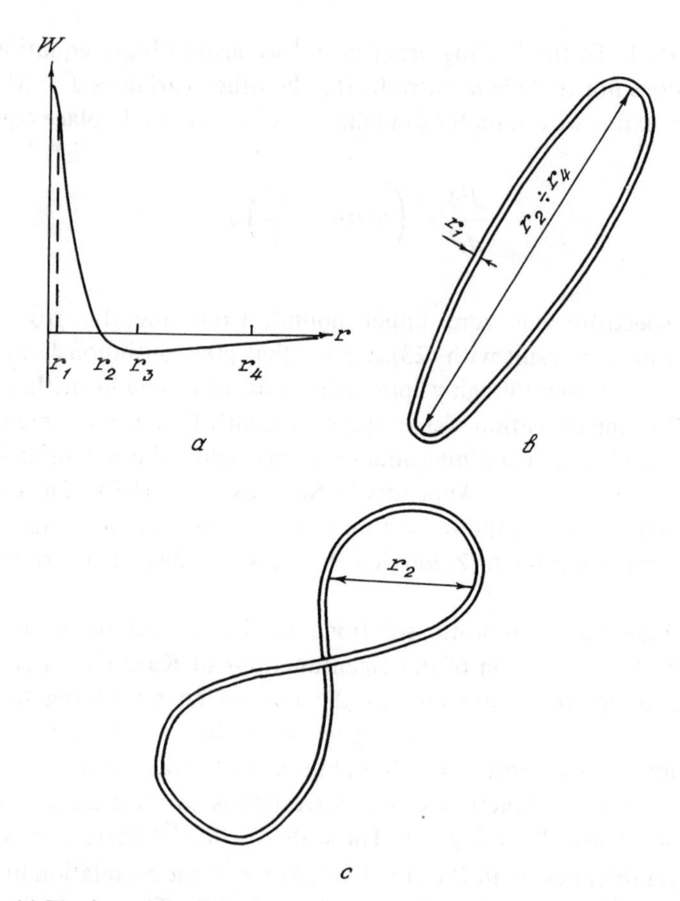

Fig. 9.5. (a) The first mode of the correlation function of the magnetic field excited by a mirror-symmetric, short-correlated flow. (b) and (c) Schematic views of the concentrations of the excited field during rope stretching (b) and twisting (c).

(closeness of the loops) which reduces to

$$\int_0^\infty Wr^2\,dr = 0$$

(note that some magnetic lines can close themselves at infinity). A certain fraction of the magnetic lines extends from the correlation cell to infinity.

This fraction can be estimated as

$$\int_0^\infty Wr^2dr \Big/ \int_0^{r_4} Wr^2dr \approx \left(1+\frac{2}{\ln r_4}\right)^{-1}.$$

The generated fluctuational magnetic field can be described as an ensemble of thin loops oriented at random and strongly elongated. The thickness of the loops is of the order of $lR_m^{-1/2}$. Between the scales $lR_m^{-1/2}$ and $lR_m^{-1/4}$, the field decreases by a factor $\ln R_m/\pi R_m^{-1/4}$ which is of the order 10^2 for the Sun for which $R_m \simeq 10^8$.

Several modes are excited simultaneously for sufficiently large magnetic Reynolds numbers. It can be easily seen that the critical R_m for the k-th mode is estimated as

$$R_m^{(k)} = \exp\left(\frac{2\pi k}{\sqrt{15}}\right),$$

i.e.,

$$R_m^{(1)} \approx 26, \quad R_m^{(2)} \approx 6.6 \times 10^2, \quad R_m^{(3)} \approx 1.7 \times 10^4.$$

Comparison with numerical calculations shows that these asymptotic values are underestimates; a more accurate estimate is $R_m^{(1)} \approx 10^2$.

When R_m is considerably higher than this critical value, the growth rates of the first two modes become very close. For instance, for $R_m = 10^8$ we obtain $(\gamma_1 - \gamma_2)/\gamma_1 \approx 0.1$. In this case the resulting correlation function of the magnetic field is a combination of the two lowest modes with comparable weights. (The third mode has a considerably lower growth rate and can probably be neglected.)

The second mode is shown in Fig. 9.6. Here $r_1 = 1.8\,R_m^{-1/2}l$, $r_2 = 0.1\,R_m^{-1/4}l$, and $r_3 = 0.6\,R_m^{-1/4}l$, while $W(r_1) \approx 0.82$, $W(r_3) = -1/8\,R_m^{-1/4}$, and $W(r_4) \sim R_m^{-5/4}$. The basic process that leads to the excitation of this mode is suggested to be the folding of doubly twisted and stretched magnetic loops shown in Fig. 9.6.

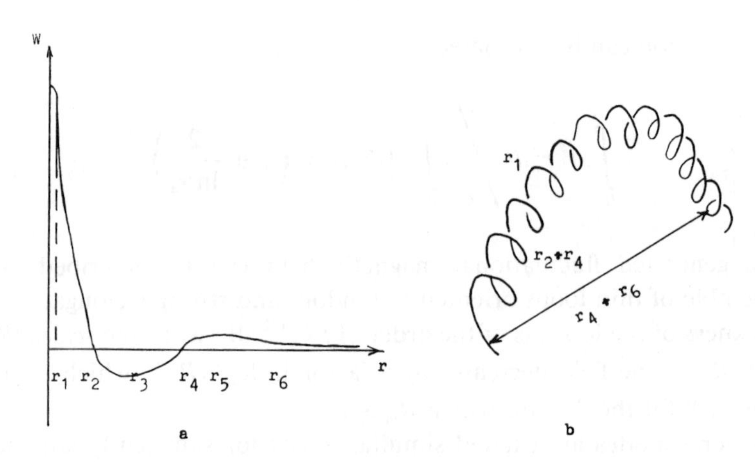

Fig. 9.6. (a) The second mode of the correlation function of the magnetic field excited by a mirror-symmetric, short-correlated flow. (b) Schematic view of the concentrations of the excited field.

Let us briefly mention the applications of these results. In the interstellar gas of our galaxy, $R_m \simeq 10^6$ (based on the ambipolar diffusion coefficient; Zeldovich *et al.*, 1983), so that $(\gamma_1 - \gamma_2)/\gamma_1 \simeq 0.2$; thus, the second mode can only be weak. Indeed, the observed correlator of the synchrotron intensity obtained by Dagkesamansky and Shutenkov (1987) strongly resembles that of the leading magnetic mode. For the Sun we have $R_m \simeq 10^8 - 10^9$ and both modes are excited with equal ease. Apart from the basic scale of turbulence, the presence of two more distinguished scales is expected, $lR_m^{-1/2}$ and $lR_m^{-1/4}$. The scale $lR_m^{-1/2}$ cannot be observed from the Earth with the modern technique: $lR_m^{-1/2} \simeq 1$ km for $l = 10^4$ km (the supergranulation scale). The scale $lR_m^{-1/4}$ corresponds to a size $\simeq 100$ km and is readily detected in observations of the solar magnetic fields (Stenflo, 1976). We stress that both observations and theory define this scale as the typical size of the regions with uniform sign of the vertical magnetic field. The structure of these fine-structure elements and the field strength in them are not connected with the mean, regular magnetic field, i.e., they should not be correlated with the phase of the solar activity cycle which is also confirmed by observations.

Our final remark has a general character. It can be shown that in the limit of weak magnetic diffusion (large magnetic Reynolds numbers) generation of small-scale magnetic fields, described in the present and previous sections, and generation of the mean (large-scale) magnetic field are independent processes. For weak helicity, the generation of small-scale magnetic fields dominates and the resulting intermittent magnetic structure is very prominent; for large helicities, the mean-field dynamo dominates and the structures inherent in this type of dynamo can be observed. The criterion in this competition is the relation between the corresponding growth rates.

9.7. THE TYPICAL STRUCTURE OF MAGNETIC FIELD

As we have emphasized above, the correlation function (the second moment) of the chaotic magnetic field generated by the fast dynamo in a random flow does not contain information on the structure of the typical distribution of the magnetic field. This is due to the fact that the principal contribution to the correlation function comes from widely spaced concentrations of the magnetic field. Thus, in order to describe a typical "background" magnetic field, we should filter out the contribution of these concentrations. This can be accomplished by analyzing the correlator of the magnetic field directions.

To describe the correlation of the magnetic field direction vectors, $e_i(\mathbf{x}, t) = H_i(\mathbf{x}, t)/H(\mathbf{x}, t)$, $i = 1, 2, 3$, we introduce the tensor

$$e_{ij}(\mathbf{x}, \mathbf{y}, t) \equiv \langle e_i(\mathbf{x}, t) e_j(\mathbf{y}, t) \rangle, \tag{24}$$

where $H(\mathbf{x}, t) = [H_i(\mathbf{x}, t) H_i(\mathbf{x}, t)]^{1/2}$ is the total field strength and the angular brackets symbolize averaging over the magnetic field distribution, which is determined here by the velocity field distribution. Due to the adopted normalization, strong maxima of the field strength introduce in (24) contributions of order unity, same as the typical background magnetic fields. However, field concentrations are rare and, therefore, the dominant contribution to (24) comes from the typical fields.

Consider the random flow of an incompressible conducting fluid whose correlation time is short when compared to the turnover time of the dominant eddies. The evolutionary equation for the tensor (24) can be derived from the general solution of the induction equation which is given in terms of the average over the random trajectories of fluid particles. Expansion of this solution in powers of Δt and substituting the result into (24) yields the desired equation for e_{ij} (this technique is discussed in detail in Section 8.6).

The resulting equation for e_{ij} is not closed because it includes higher moments of the direction vector (up to the sixth moment). In order to close the equation, we adopt a simplifying assumption that, like the moment of a Gaussian random field, these higher correlators can be represented as the products of various pair-wise correlators. This leads to a closed equation for the correlator e_{ij} which is given here for the case where the stationary solution (an analogue of the eigenfunction) is isotropic:

$$e_{ij}(r) = \frac{1}{3}\left[c(r)\delta_{ij} + s(r)\frac{r_i r_j}{r^2}\right], \qquad (25)$$

where the desired functions $c(r)$ and $s(r)$ decay at infinity, $c(0) = 1$ and $s(0) = 0$. Notice that in contrast to the velocity and the magnetic fields, the vector e_i is not solenoidal, so that the correlator (25) cannot be described by a single scalar function. Thus, we have

$$\begin{aligned}
&(1 + \varepsilon - F)c'' + \left[-2F' + \frac{2}{r}(\varepsilon + 1 - F) + \frac{4}{5}\frac{F_2}{r}\right]c' \\
&+ \left[-\frac{3}{5}\left(F_1'' + \frac{2}{r}F_1' + 2\frac{F_2}{r^2}\right) - \frac{1}{35}\left(F_1'' + \frac{5}{r}F_1' - \frac{1}{r}F_2'\right.\right. \\
&\left.\left. + 2\frac{F_2}{r^2}\right) - 4F''(0)\right]c + \left[-F_1'' + \frac{2}{r}F_1' - 2\frac{F_1}{r^2} + (\varepsilon + 1)\frac{2}{r^2}\right. \\
&\left. + \frac{2}{5}\left(F_1'' + \frac{1}{r}F_1'\right) - \frac{1}{35}\left(F_1'' + \frac{1}{r}F_1' + \frac{3}{r}F_2' - \frac{4F_2}{r^2}\right)\right]s \\
&- \frac{1}{35}\left[4F_1'' + F_2'' + \frac{11}{r}F_1' + \frac{3}{r}F_2' + \frac{2F_2}{r^2}\right]c^3 \\
&- \frac{1}{35}\left[5F_1'' + 2F_2'' + \frac{3}{r}F_1' + \frac{5}{r}F_2' - 8\frac{F_2}{r^2}\right]c^2 s \\
&- \frac{2}{35}[F_1'' + F_2'']s^2 c = 0, \qquad (26)
\end{aligned}$$

$$
\begin{aligned}
&(1+\varepsilon-F)s''+\left(\frac{21}{5}F'\right)c'+\left[\frac{2}{r}(\varepsilon+1-F)+\frac{9}{5}F'\right]s' \\
&+\left[-\frac{3}{5}\left(F_2''+\frac{2}{r}F_2'-\frac{6F_2}{r^2}\right)-\frac{1}{35}\left(3F_1''+2F_2''+\frac{F_2'}{r}-\frac{6F_2}{r^2}\right)\right]c \\
&+\left[-F_2'+\frac{4F_2}{r^2}-\frac{2}{r}F_1'+6\frac{F_1}{r^2}+(\varepsilon+1)\frac{6}{r^2}\right. \\
&+\frac{2}{5}\left(F_1''+2F_2''-\frac{1}{r}F_2'-2\frac{F_2}{r^2}+\frac{F_1'}{r}\right) \\
&-\frac{1}{35}\left(3F_1''+4F_2''-\frac{3}{r}F_2'+\frac{4F_2}{r^2}\right) \\
&\left.-4F'''(0)\right]s-\frac{6}{35}[F_1''+F_2'']s^3 \\
&-\frac{1}{35}\left(2F_2'+3F_1''+\frac{F_2'}{r}-\frac{3}{r}F_1'-\frac{6}{r^2}F_2\right)c^3 \\
&-\left[14F_1''+9F_2''+\frac{9}{r}F_1'-\frac{3}{r}F_2'-\frac{4F_2}{r^2}\right]\frac{c^2s}{35} \\
&-\left[16F_1''+12F_2''+\frac{4}{r}F_2'-\frac{8}{r}F_2\right]\frac{cs^2}{35}=0,
\end{aligned}
\tag{27}
$$

where $\varepsilon = 6\nu_m/(l\nu)$ is inversely proportional to the magnetic Reynolds number, the prime denotes derivatives with respect to r, all functions being taken at the point r, $F_1 = F + F'/2$ and $F_2 = -r/2F'$.

We analyze the nonlinear system of the equations (26), (27) in the approximation of the small ε, i.e., large R_m, using the asymptotic method applied to the magnetic correlator in the last section. Let us put $F(r) = \exp(-r^2)$ with r measured in units of l and consider separately the following three regions: $0 < r \leqslant a\varepsilon^{1/2}$, $a\varepsilon^{1/2} < r \leqslant b$ and $b < r$, where a and b are constants of the order of unity which will be determined in the course of the solution.

In the first region, we introduce the variable $\xi = r\varepsilon^{-1/2}$. Retaining only the leading terms in ε in (26), (27), we have

$$\begin{aligned}
&(1+\xi^2)\ddot{c} + \left[\frac{2}{\xi}(1+\xi^2) + \frac{24}{5}\xi\right]\dot{c} + \frac{102}{7}c \\
&\quad + \frac{50}{35}c^3 + \left(\frac{2}{\xi^2} - \frac{78}{35}\right)s = 0, \\
&(1+\xi^2)\ddot{s} + \frac{42\xi}{5}\dot{c} - \left[\frac{6}{\xi^2} + \frac{124}{35}\right]s \\
&\quad + \left[\frac{2}{\xi}(1+\xi^2) - \frac{18}{5}\xi\right]\dot{s} = 0,
\end{aligned} \tag{28}$$

where the dots denote derivatives with respect to ξ. The system (28) has a one-parametric family of solutions which satisfy the boundary conditions $c(0) = 1$ and $s(0) = 0$. This family can be obtained, e.g., through expansions in powers of ξ. The first terms of these series are given by

$$\begin{aligned}
c(\xi) &= 1 - \frac{8 + C_1}{3}\xi^2 + \dots \\
s(\xi) &= C_1\xi^2 + \dots .
\end{aligned} \tag{29}$$

In the second region we introduce the variable $\zeta = r\varepsilon^{-1/4}$. Again, keeping in (26), (27) only the leading terms in ζ, we have

$$\begin{aligned}
&\zeta^2\ddot{c} + \frac{34}{5}\zeta\dot{c} + \frac{102}{7}c - \frac{78}{35}s = 0, \\
&\zeta^2\ddot{s} + \frac{8}{5}\zeta\dot{s} - \frac{42}{5}\zeta\dot{c} + \frac{156}{35}s = 0,
\end{aligned} \tag{30}$$

where the dots denote derivatives with respect to ζ. When deriving these equations, we expect that in the second region the desired solution is small as compared with $c(0) = 1$ so that we can neglect the nonlinear terms. Expressing s through c from the first equation of (30), we obtain the fourth-order Euler equation for c which has the following four-parameter solution:

$$c(\zeta) = \zeta^{\alpha}[C_2 \cos(\beta \ln \zeta) + C_3 \sin(\beta \ln \zeta)] + \zeta^{\gamma}[C_4 \cos(\delta \ln \zeta) + C_5 \sin(\delta \ln \zeta)], \tag{31}$$

where the constants α, β, γ and δ can be obtained by numerical solution of the fourth-order characteristic equation $\alpha = 2.35$, $\beta = 2.08$, $\gamma = -3.62$ and $\delta = 3.36$.

In the third region, $r > l$ in (26), (27) and we can put $\varepsilon = 0$. The resulting solution can be easily analyzed for $\tau \gg 1$, when (26), (27) reduce to

$$c'' + \frac{2}{r}c' + 8c + \frac{2s}{r^2} = 0,$$

$$s'' + \frac{2}{r}s' + 8s + \frac{6c}{r^2} = 0.$$

The only stationary solution of this system of equations is a four-parametric family of solutions which decay at infinity slower than r^{-1}. This slow decay of the correlator probably means that for very large r the stationary approximation (26), (27) is inapplicable and the complete non-stationary problem should be considered, which describes the propagation of the correlator front.

In order to obtain the complete asymptotic solution, all these asymptotic forms at the boundaries of all three regions should match. In the framework of the adopted technique, the matching is carried out to two successive orders in ε. The leading order asymptotic allows us to match the functions c and s, while matching of their derivatives is accomplished with the help of the second-order solution. Such two-step matching is necessary because the derivatives of the solution (31) have a higher order in ε than the functions themselves. As shown in Section 9.6, qualitatively correct results can be obtained only when the functions are matched and the values $a = b = 1$ are chosen while s vanishes both at the boundary between the first and second regions and at the origin. At the boundary between the second and third regions, we put $c = s = 0$ in order to minimize the slowly decaying tail of the correlator. Here we restrict ourselves to this crude approximation considering, for simplicity, expressions (29) as exact. Keeping only the leading terms in ε, the result reads

$$c(r) = \begin{cases} 1 - \dfrac{r^2}{\varepsilon}, & 0 < r < \sqrt{\varepsilon}, \\ A \dfrac{r^8}{\varepsilon^{\gamma/2}} \left[\cos\left(\delta \ln \dfrac{r}{\varepsilon^{1/4}}\right) - \sin\left(\delta \ln \dfrac{r}{\varepsilon^{1/4}}\right), \right. & \sqrt{\varepsilon} < r \leqslant 1, \\ 0 & r \geqslant 1, \end{cases}$$

$$s(r) = \begin{cases} 0 & 0 < r \leqslant \sqrt{\varepsilon}, \quad r \geqslant 1 \\ -A \dfrac{r^8}{\varepsilon^{\gamma/2}} \left[8.2 \cos\left(\delta \ln \dfrac{r}{\varepsilon^{1/4}}\right) + 9.3 \sin\left(\delta \ln \dfrac{r}{\varepsilon^{1/4}}\right) \right], & \sqrt{\varepsilon} < r \leqslant 1 \end{cases} \tag{32}$$

$$A = -\frac{5}{3} \frac{1}{\cos(\ln \varepsilon^{1/4}) - \sin(\ln \varepsilon^{1/4})}.$$

These functions are shown schematically in Fig. 9.7. Comparing the solution (32) with the magnetic field correlator obtained in a similar approximation in Section 9.6, we see that they are rather similar. By order of magnitude, the maxima near the origin have equal widths which indicates that even the weaker background magnetic field is concentrated within the ropes and, possibly, skin-layers of thickness $\sim \sqrt{\varepsilon}$. The shape of the ropes (and layers) is determined by the behaviour of the correlator for $\varepsilon^{1/2} \leqslant r/l \leqslant 1$. The correlator (32) has $\sim \ln R_m$ zeros within this range, which can be interpreted as an indication of the presence of multi-twisted ropes (see Section 9.6) with the twisting number $\sim \ln R_m$. Note that the higher peaks that contain the major portion of the magnetic energy have smaller twisting numbers of the order of unity.

Analyzing the derivation of (32), we notice that the specific closure hypothesis for the equations (26), (27) is essential only for the solution in

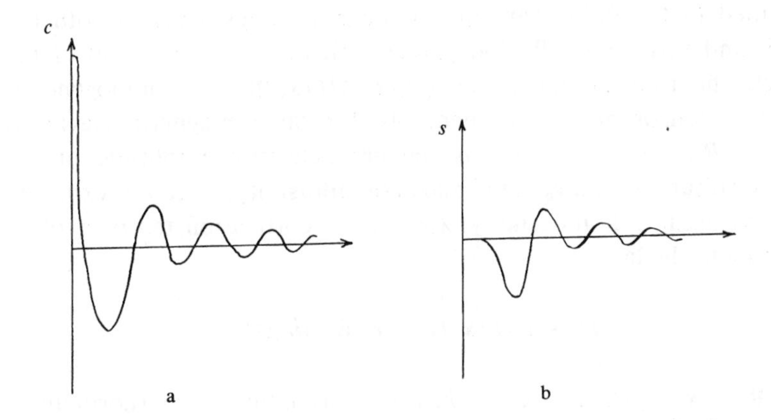

Fig. 9.7. The schematic form of the functions $c(r)$ (a) and $s(r)$ (b).

the first region which describes the field distribution across a layer or rope (the nonlinear terms that arise due to the closure of the equations (26), (27) are essential only in this region). However, in practically interesting cases of solar or galactic magnetic fields, these scales are so small that the correlators c and s will hardly observable at these scales in the foreseeable future. On the contrary, the scales associated with the second region are in principle observable (see Section 9.6).

9.8. SELF-EXCITATION OF MAGNETIC FIELDS IN FINITE BODIES

Above we have considered self-excitation of magnetic fields in spatially unbounded, statistically homogeneous flows. Real flows are confined to finite volumes. This introduces an additional effect that opposes field generation: the field diffusionally leaks through the volume boundaries, which leads to the additional excitation condition of sufficiently large size of the volume as compared with the turbulent correlation scale. In this section, we will consider this effect for the case of self-excitation of fluctuative magnetic fields (Maslov *et al.*, 1987).

Consider the turbulent flow of a conducting fluid in a volume of characteristic size L such that $\lambda = l/L \ll 1$, where l is the correlation scale of the velocity field. Suppose that some initial weak field is embedded in the fluid. Tangling of magnetic lines leads to both the growth and turbulent diffusion (along with the molecular one) of the magnetic field correlation tensor $\langle H_i(\mathbf{x}, t) H_j(\mathbf{x}, t)\rangle$. In a homogeneous flow, this tensor grows exponentially for large magnetic Reynolds numbers $R_m = lv/\nu_m$ (here v is the characteristic amplitude of the velocity pulsations and ν_m is the magnetic diffusivity). It can be expected that for a weak inhomogeneity, $\lambda \ll 1$, the correlation tensor evolves according to the law

$$\langle H_i(\mathbf{x}, t) H_j(\mathbf{x}, t)\rangle = E(\mathbf{R}, t) h_{ij}(\mathbf{r}),$$

where $\mathbf{R} = (\mathbf{x} + \mathbf{y})/2$, $\mathbf{r} = \mathbf{x} - \mathbf{y}$, h_{ij} is the correlation tensor (normalized to unity: $h_{ij}(0) = \delta_{ij}$) for the corresponding homogeneous flow, the quantity E is proportional to the magnetic energy density at position R. The latter can be shown to obey the following Schrödinger-type equation:

$$\Gamma E = \gamma(\mathbf{R}) E + \lambda^2 D(\mathbf{R}) \Delta E \tag{33}$$

written in dimensionless variables with the distance R measured in units of l and the time in units of l/v, where γ is the growth rate of the accompanying homogeneous flow problem, Γ is the global (true) growth rate, and D is the turbulent diffusivity of magnetic energy. This equation was proposed by Novikov *et al.* (1983). It can be derived from the equation of the second moment of magnetic field in short-correlated flow for $\lambda \ll 1$. It is presumed that the flow is locally isotropic and reflectionally symmetric. The spatial part of the correlation tensor for such flow can be represented as

$$\langle v_i(\mathbf{x}) v_j(\mathbf{y})\rangle = \varphi(\mathbf{R}) \frac{V_{ij}(\mathbf{r})}{3} + O(\lambda),$$

where V_{ij} is the isotropic part of the tensor while the function $\varphi(R)$ is bounded and non-vanishing within the considered volume, and vanishes outside. With these approximations, for $R_m \gg 1$ we have the following asymptotic relations (see Section 9.6):

$$\gamma(\mathbf{R}) = \varphi(\mathbf{R})\left(\frac{3}{4} - \frac{4\pi^2}{5}(\ln R_m)^{-2} + \dots\right),$$
$$D(\mathbf{R}) = \frac{1}{6}\varphi(\mathbf{R}) + \frac{1}{2}R_m^{-1}. \tag{34}$$

Equation (33), supplemented by the condition for the continuity of $E(\mathbf{R})$ at the boundary and finiteness of $E(\mathbf{R})$ at infinity, represents a boundary-value problem for the eigenvalues r_n and the eigenfunctions E_n. The smallness of λ allows us to apply the WKB approximation.

Consider the simple cases of spherically symmetric, cylindrically symmetric and flat distributions of the velocity field when the function $\varphi(\mathbf{R})$ is defined as

$$\varphi(\mathbf{R}) = \begin{cases} 1 - R^2 & |\mathbf{R}| < 1 \\ 0 & |\mathbf{R}| > 1 \end{cases}$$

where R is the distance from the origin, the axis, or the symmetry plane, respectively. Then the lower (i.e., corresponding to small values of n) eigenvalues, i.e., the growth rates of magnetic energy, are given by

$$\Gamma_n = \gamma(0) - \lambda(6R_m^{-1} + 2)^{1/2}(n + 3/4),$$
$$\Gamma_n = \gamma(0) - \lambda(6R_m^{-1} + 2)^{1/2}(n + 1/2), \tag{35}$$
$$\Gamma_n = \gamma(0) - \frac{\lambda}{2}(6R_m^{-1} + 2)^{1/2}(n + 1/2),$$

where $n = 0, 1, 2, \dots$ and $\gamma(0)$ are the homogeneous-flow growth rates given by (34). The field is self-excited, i.e., the growth rate Γ_n is positive when R_m is sufficiently large and λ is sufficiently small. In the limit $\lambda \to 0$, the growth rates Γ_n tend to $\gamma(0)$. Inhomogeneity leads to the appearance of the generation threshold in λ: for a given R_m the growth rate is positive when $\lambda > \lambda_{cr}$ with

$$\lambda_{cr} = \frac{4}{3}\gamma(6R_m^{-1} + 2)^{-1/2},$$

$$\lambda_{cr} = 2\gamma(6R_m^{-1} + 2)^{-1/2}, \tag{36}$$

$$\lambda_{cr} = 4\gamma(6R_m^{-1} + 2)^{-1/2},$$

respectively. We see that for values of R_m close to the generation threshold for homogeneous flows, the resulting values of λ_{cr} are indeed small. The case of planar symmetry turns out to be most favourable for generation. For large R_m and small λ the eigenfunction E_0 concentrates near the maximum of $\varphi(\mathbf{R})$. Note that in the subcritical regimes, $R_m < (R_m)_{cr}$, the magnetic energy does not decay monotonically. For sufficiently large magnetic Reynolds numbers, rather prolonged stages of the temporal growth of the magnetic energy occur (Zeldovich, 1956; Maslova and Ruzmaikin, 1987).

9.9. MAGNETIC FIELD INTERMITTENCY WITH ACCOUNT OF THE LORENTZ FORCE

When the magnetic field grows exponentially in a random flow, the approximation of fixed velocity field (the kinematic dynamo) becomes inapplicable after a rather short period. The Lorentz force produced by the magnetic field on the flow becomes essential. Obviously, this back-action leads to considerable effects primarily on the field concentrations and the Le Chatelier principle suggests that this results in smoothing of the intermittency. Accounting for this effect makes the picture of intermittency more realistic; even though magnetic field concentrations in nature are strong, they never have such catastrophic character as predicted by the linear model of intermittency. However, one may wonder whether or not the nonlinearity can remove all the structural details resulting from random flow at the linear stage.

In this section we make an attempt to describe how the magnetic field influence on the flow affects the intermittency. To this end, we introduce into the induction equation the nonlinearity that describes this back-action. Our principal result is the assertion that the steady (limiting) magnetic field distribution preserves intermittency, i.e., it remains to be structured under certain, not very restrictive, conditions. In a short time interval, the magnetic field frozen into a fluid particle changes according to the following law that follows from the induction equation:

$$H_i(t + \Delta t) = \left(\delta_{ij} + \frac{\partial v_i}{\partial x_j}\Delta t\right) H_j(t), \quad i, j = 1, 2, 3. \tag{37}$$

Suppose that the mean velocity vanishes and the velocity field is a short-correlated, homogeneous and isotropic random process. This is equivalent to every element of the matrix $\partial v_i/\partial x_j \, \Delta t$ being a Brownian (or Wiener) random process with zero average value and certain dispersion σ^2, which can be represented as

$$\frac{\partial v_i}{\partial x_j}\Delta t = \sigma\left(w_{ij} - \frac{1}{3} w_{ll}\,\delta_{ij}\right),$$

for an incompressible flow, where w_{ij} are independent Wiener processes with unit dispersion and w_{ll} is the trace of the matrix w_{ij}.

Equation (37) does account for the back-action of the magnetic field on the velocity field and diffusion. These effects can be described phenomenologically by adding on the right-hand side of (37) the term $-\eta H_i(t)$ for magnetic diffusion and by multiplying the matrix $\partial v_i/\partial x_j$ by a nonlinear function $g(H)$ for the back-action, so that $g \to 0$ for $H \to \infty$. A particular example of a suitable nonlinear function is given by $[1 + (H/H_*)^2]^{-1}$, where H_* is the characteristic field strength above which nonlinear effects become pronounced. Thus, we consider the following model of nonlinear evolution of the magnetic field:

$$H_i(t + \Delta t) = H_i(t) + g(H)\sigma\left(w_{ij} - \frac{1}{3} w_{ll}\,\delta_{ij}\right) H_j(t) - \eta H_i(t). \tag{38}$$

In order to derive the equation for the field strength, we raise this equation to the second power. Taking into account that linear combinations of the type $w_{ij}H_iH_j/H^2$ are also Wiener processes with zero average while the products $w_{ij}w_{ik}$ after averaging become proportional to Δt (see

Section 9.5), we obtain the following equation for the field strength:

$$dH = \frac{4}{3}\sigma^2 g^2 H\,dt + \left(\frac{2}{3}\right)^{1/2} \sigma g\, H\,dw_t - \eta H\,dt, \tag{39}$$

where w_t is the Wiener process with zero mean value and unit dispersion.

The field strength H in (3) is a random process. When H is small and $g \sim 1$, the field grows exponentially (Section 9.5). We expect that the nonlinearity and diffusion saturate this growth and a certain steady state distribution is eventually established.

The random process described by (39) can be considered as one-dimensional diffusion along the H-axis with diffusivity $D \equiv \frac{1}{3}\sigma^2 g^2(H)H^2$ accompanied by advection along this axis with velocity $V = \frac{4}{3}\sigma^2 g^2 H - \eta H + \frac{1}{3} g g' H^2$ (cf. Section 8.9). This combined process corresponds to the following operator:

$$\hat{L} = D\frac{d^2}{dH^2} + V\frac{d}{dH}. \tag{40}$$

The desired steady-state probability density of the magnetic field $p(H)$ is the limit for $t \to \infty$ of the transitional probability density $\Pi(t, \tilde{H}, H)$ for the transition from the field $\tilde{H}$ to the field H in time t. Notice that the limiting probability density does not depend on the initial state $\tilde{H}$. The transitional probability density obeys the following equation:

$$\frac{\partial \Pi}{\partial t} = \hat{L}(\tilde{H})\Pi = \hat{L}^*(H)\Pi$$

and the initial condition $\Pi(0, \tilde{H}, H) = \delta(\tilde{H} - H)$. The asterisk denotes the adjoint operator while the second equality is due to the time-independence of $\hat{L}$, i.e., to the symmetry with respect to time inversion, $t \to -t$. Passing on to the limit $t \to \infty$ when $\partial\Pi/\partial t = 0$, we obtain the following equation for the limiting probability density:

$$\hat{L}^*(H)p(H) = 0. \tag{41}$$

Of course, the probability density is normalized by $\int_0^\infty p dH = 1$.

Substituting the operator

$$\hat{L}^* p = \frac{d^2}{dH^2}(Dp) - \frac{d}{dH}(Vp)$$

into (41) and solving this equation, we obtain the desired limiting probability density for the field strength H:

$$p(H) = C\frac{H^2}{g(H)} \exp\left(-\frac{3\eta}{\sigma^2}\int \frac{dH}{Hg^2(H)}\right). \tag{42}$$

One of the integration constants is equated to zero as required by the finiteness of the integral $\int p dH$ and the remaining constant C follows from the normalization condition $\int p dH = 1$.

Thus, for $g = [1 + (H/H_*)^2]^{-1}$ we obtain

$$p = C\left[1 + \left(\frac{H}{H_*}\right)^2\right] H^{2-3\eta/\sigma^2} \exp\left[\frac{3\eta}{\sigma^2}\left(\frac{H}{H_*}\right)^2 - \frac{3\eta}{4\sigma^2}\left(\frac{H}{H_*}\right)^4\right].$$

The following simple approximate expression can also be used:

$$p(H) = \begin{cases} 3H_*^{-1}\left(1 - \dfrac{\eta}{\sigma^2}\right)\left(\dfrac{H}{H_*}\right)^{2-\frac{3\eta}{\sigma^2}}, & H < H_* \\ 0 & , \quad H > H_*. \end{cases} \tag{43}$$

This expression as well as similar expressions previously given are valid for $\eta/\sigma^2 < 1$. Otherwise, the distribution (42) cannot be normalized and the limiting distribution is $\delta(H)$, i.e., for $t \to \infty$ the field strength tends to zero.

The form of the limiting distribution (43) suggests that the probability density is a decreasing function of H for $1 > \eta/\sigma^2 > 2/3$, i.e., typical limiting strength of the magnetic field is much weaker than H_* and the most probable field strength is zero. For $\eta/\sigma^2 < 2/3$, the probability density grows with H and most probably $H \simeq H_*$.

The mean energy density of the magnetic field is proportional to $\int H^2 p dH \sim H_*^2$ in both the cases of the previous paragraph. This implies that for $2/3 < \eta/\sigma^2 < 1$, the limiting distribution is intermittent: the peaks with $H \simeq H_*$, which determine the mean energy, stand against a weak background of the strength $H << H_*$. A peculiar field distribution arises for $\eta/\sigma^2 < 2/3$: the widely scattered regions with weak field in a rather homogeneously strong field distribution. This can be verified by calculating the statistical moments $\langle |H|^{-q} \rangle$ which diverge for certain $q > 0$. Of course, such "holes" are present also for larger values of η/σ^2 but they are less pronounced because the typical field amplitude is smaller.

The nonlinear model (38) is only a crude approximation to a consistent nonlinear problem. In particular, this model does not take into account that nonlinear effects (magnetic force and associated velocity perturbations and magnetic field influence on magnetic diffusivity) depend not only on the strength of the magnetic field but also on the field direction. In addition, replacement of the magnetic diffusion $\nu_m \Delta H$ by friction $-\eta H$ is not a very accurate procedure. An adequate allowance for magnetic diffusion would probably result in the vanishing of $p(H)$ and its derivatives for $H \to 0$. Then for large $\eta/\sigma^2 (>> 1)$ the probability density $p(H)$ would be concentrated at small H, i.e., a typical magnetic field would be weak. For moderate values of η/σ^2 (from 2/3 to 1) the singularity of $p(H)$ at $H = 0$ would be smoothed out and the most probable value of H would be small but finite. For small values of $\eta/\sigma^2 (<2/3)$ $p(H)$ and all its derivatives would vanish for $H \to 0$ so that the mean values $\langle |H|^{-q} \rangle$ would be large but finite. In physical terms, these properties imply that the magnetic fields spread into the holes in the field distribution diffusionally.

It is, however, remarkable that even the simplest model (38) reflects some realistic properties of the nonlinear dynamo.

Indeed, the parameter η/σ^2 can be understood as an inverse magnetic Reynolds number for a certain scale of field concentration. Obviously, the magnetic Reynolds number is of the order of unity for the scales near dissipation. Therefore, this result can be interpreted as an indication of the magnetic field intermittency at small scales. On the other hand, we can also understand this result as an indication of intermittency for the moderate magnetic Reynolds numbers which are characteristic, e.g., of the threshold for field self-excitation; this situation is analogous to the hydrodynamic transition from laminar to turbulent flow, during which flow structures arise.

9.10. THE DYNAMO THEOREM

The picture of evolution of the magnetic field in a short-correlated flow of a conducting medium outlined above is prescribed for other types of stochastic flows with finite memory time. For instance, in a flow with loss of memory of fixed moments τ, 2τ, 3τ, ... (the renovating flow) a typical realization of the magnetic field grows exponentially while the mean values of the consequent power of the magnetic field strength have progressively increasing growth rates, so that intermittent field distributions are established. This statement is the main content of the dynamo theorem of Molchanov *et al.* (1984). However, the finite renovation time evaluation of the growth rate is complicated by the difficulties in calculating the magnetic field increments for the finite time increment of the order τ. These increments are determined by the evolutionary Green's matrix, so that the field values at the moments $n\tau$ and $(n + 1)\tau$ are related by

$$H_i[(n + 1)\tau] = G_{ij}H_j(n\tau)$$

along the Lagrangian path. The evolution of the field after many renovation cycles is described by the product of the corresponding Green's matrices G_{ij}, which are all independent because the flow is renovating. The problem of evaluating the product of a large number of independent random matrices was first solved by Furstenberg (1963). The mathematical formalism used in proving the dynamo theorem is rather complicated. Here we restrict ourselves to a discussion of the result while the detailed presentation is given by Molchanov *et al.* (1984, 1985).

At first sight, symmetry properties suggest that the product of the independent random matrices must tend, in the isotropic case, to the unit matrix multiplied by the mean value of the determinants of the multipliers. Actually, a more delicate phenomenon occurs: a spontaneous violation of symmetry owing to which a certain random set of basis vectors is singled out that determines the directions of the magnetic vector fields which either grow or decay without strong change of their directions. In order to determine how a given magnetic field vector would evolve, it is sufficient to decompose it into three components directed along the random basis vectors. These components change at different rates so that eventually the total field vector approaches in the direction of that basis vector along which the field grows most rapidly (apart from the degenerate case where the initial field is orthogonal to this direction). The randomness of this set of basis vectors means that this set is formed during the evolution of the magnetic field and one can determine in advance only the correlation properties of the basis vectors. In contrast, the growth rates of the magnetic field components directed along the basis vectors are deterministic and they can be determined in advance. The concept of spontaneous violation of the symmetry of the products of random matrices can be traced back to the ideas of Lyapunov whose name is immortalized in Lyapunov's exponents (which are essentially the growth rates of the field components) and Lyapunov's random basis. To some extent, other ideas of the intermittency theory also date back to Lyapunov (in particular, this is the idea of difference between growth rates of consequent statistical moments). Unfortunately, the development of these ideas experienced a lull that lasted for more than half a century after the tragic death of Lyapunov during the revolt in Odessa in 1919. Remaining alive in the mathematical consciousness, these ideas were rediscovered only later.

In order to understand the result of consecutive actions of a large number of independent random matrices on a given vector $\mathbf{H}_0$ qualitatively, imagine a two-dimensional sphere whose radius is equal to the length of this vector. Under the action of a single matrix $\hat{G}_1$ with unit determinant, the sphere is transformed into an ellipsoid of the same volume. Let the extension factor along the x-axis be λ. Then the conservation of the volume (or area in two dimensions) requires that along the y-axis contraction occurs by a factor λ^{-1}. After this transformation, the vector length becomes

$$(\lambda^2 H_{0x}^2 + H_{0y}^2 \lambda^{-2})^{1/2} = H_0(\lambda^2 \sin^2\varphi + \lambda^{-2}\cos^2\varphi)^{1/2},$$

where φ is the polar angle. When $\cos\varphi > (1+\lambda^2)^{-1/2}$, this length exceeds the initial one, $H_0 = (H_{0x}^2 + H_{0y}^2)^{1/2}$. This shows that the set of directions φ, for which $\mathbf{H}_0$ is stretched to become larger, is more than half of all possible directions. In other words, an arbitrary vector is stretched with a probability exceeding 1/2. Therefore, the action of the large number of matrices $\hat{G}$ with independent directions of stretchings and independent stretching coefficients leads to net growth of the vector length.

The unit determinant of the matrices G corresponds to the incompressibility of the fluid in which the magnetic field is embedded. It is not difficult to take into account the difference of the determinants from unity because the determinants of the multiplier matrices, which are simply scalar factors, commute with all the matrices and can be multiplied separately.

Hence, practically any magnetic field vector eventually approaches in the direction leading to the vector of the Lyapunov basis and its growth rate becomes

$$\gamma^{(1)} = \frac{1}{\tau} \ln g,$$

where $g = \langle |\hat{G}\mathbf{H}_0|/|H_0| \rangle$ is the mean magnitude of the stretching by a single matrix, and τ is the renovation time. Those degenerate vectors which are orthogonal to one or even two of the vectors of Lyapunov's basis evolve at smaller rates $\gamma^{(2)}$ and $\gamma^{(3)}$. Due to the fact that the determinants of the matrices $\hat{G}$ are unity, these growth rates are related by

$$\gamma^{(1)} + \gamma^{(2)} + \gamma^{(3)} = 0.$$

Common sense suggests that this relation implies, e.g., $\gamma^{(1)} > 0$ and $\gamma^{(3)} < 0$ because it is hardly probable that three more or less arbitrary numbers vanish simultaneously. Nevertheless, this conjecture, which singles out possible degenerate situations when all the three Lyapunov exponents vanish, is very difficult to prove rigorously. It is a curious oddity that one first shows the presence of intermittency, i.e., difference of growth rates of different moments, and only then the positiveness of one of the Lyapunov exponents, e.g., $\gamma^{(1)}$, follows.

From the standpoint of the probability theory, intermittency appears as a consequence of the so-called Lyapunov inequality for statistical moments, which states that for every random quantity the following inequality holds:

$$\langle \xi^{p+q} \rangle \geqslant \langle \xi^p \rangle \langle \xi^q \rangle .$$

In turn, this inequality is similar to the well known inequality relation between the arithmetic average value and the geometric average value. However, this approach only proves the presence of intermittency, leaving out the obscure specific processes of evolution of magnetic lines which lead to the intermittency. This latter question was addressed by Finn and Ott (1988) who were involved in the mapping theory which is widely used in the studies of chaos in dynamic systems. In the framework

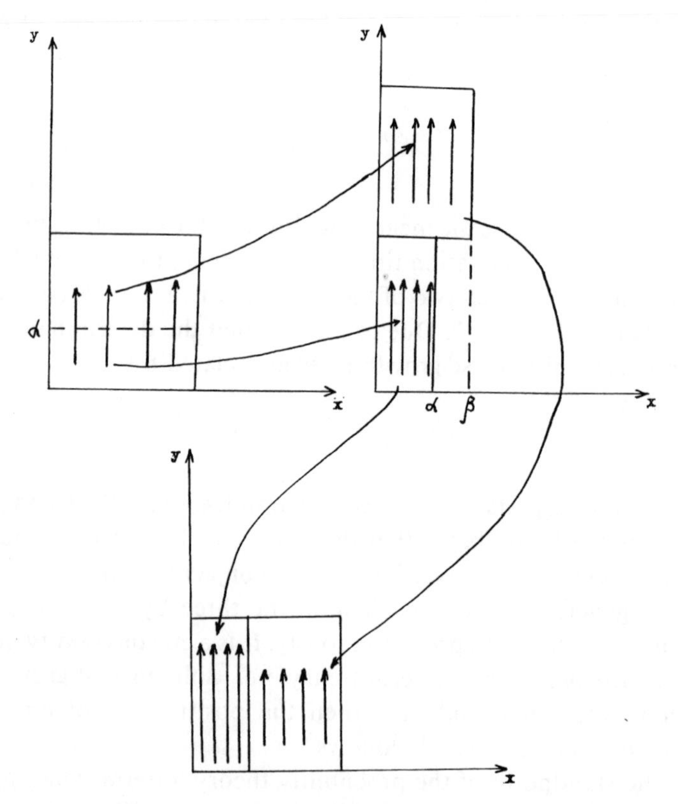

Fig. 9.8. The "figure-eight" rope dynamo in terms of mappings.

of this formalism, we replace a detailed flow in the "figure-eight" dynamo of Fig. 9.2 by the following discrete mapping (Fig. 9.8). Consider the magnetic loop initially situated on the plane (x, y) and deformed according to the law:

$$x_{n+1} = \begin{cases} \alpha x_n & y_n < \alpha, \\ \beta x_n + \alpha & y_n > \alpha, \end{cases}$$
$$y_{n+1} = \begin{cases} y_n/\alpha & y_n < \alpha, \\ (y_n + \alpha)/\beta & y_n > \alpha, \end{cases} \tag{44}$$

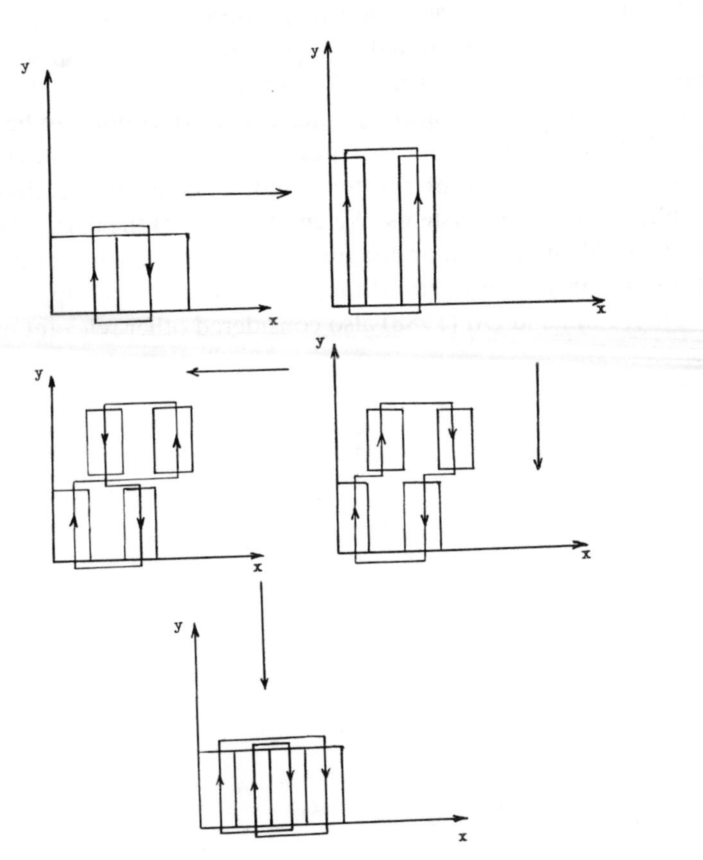

Fig. 9.9. The "figure-eight" dynamo can be also realized as a series of continuous deformations.

where x and y are coordinates of the points of the loop and n is the mapping cycle number. It can be easily verified that this type of deformation leads to the same final configuration as the sequence of continuous deformations shown in Fig. 9.9. The mapping (44) occurs on two-dimensional manifold and is discontinuous; however, it leads to the same result as continuous three-dimensional transformation which can hardly be described analytically.

The mapping (44) can be analyzed with the help of the standard methods of the mapping theory. It turns out that the field strength in the magnetic loop transformed by this mapping becomes intensified and the corresponding phase-space trajectory approaches a strange attractor whose dimension is fractional and smaller than the dimension of the phase space. This implies that the growing magnetic field occupies a small fraction of the physical volume. This distribution can be naturally identified with the intermittent peaks of the magnetic field. A more realistic description of intermittency requires the introduction of more complicated dynamic systems. We emphasize that the particular form of mapping (44) is not important for the presence of intermittency, and determines only the attractor dimension and the rate of approach to the attractor. Finn and Ott (1988) also considered other relevant mappings.

CHAPTER

TEN

THE CASUAL UNIVERSE

The role of chance and the formation of structures associated with randomness manifest themselves in very diverse physical processes in our universe. Recent developments have put forward the possibility of a very basic role of chance in the universe: the universe itself can be a result of casualness. To some extent, these notions remind us of the ideas of L. Boltzmann who considered the surrounding world an enormous fluctuation. This chapter discusses the most universal role of chance following a review of Zeldovich (1988).

10.1. THE UNIVERSE

The size of the universe, and even the size of the explored part of the universe, to be more moderate and exact, by far exceeds the possibilities of human imagination.

In ancient times, people could hardly agree that the Earth is a sphere. Nowadays, when airplanes cover thousands of kilometers without landing, in the age of spaceflights, radio and television (and, unfortunately, in the age of inter-continental missiles with nuclear warheads) the Earth has become a small fragile ball. Now, nobody is much impressed by the distance of the Earth from the Sun, which is 150 million kilometers, the so-called astronomical unit. However, the distance of the solar system to the center of the Galaxy (about 10 kiloparsecs = 3×10^{22} cm) exceeds the distance between the Earth and the Sun by a factor of two billion. Furthermore, the greatest distance at which the brightest galaxies are observable is of the order of a few thousand megaparsecs, larger by a factor of a million than the distance from the

Sun to the Galactic centre. This distance is greater than the astronomical unit by the same factor as the astronomical unit exceeds a small dust particle, whose size is far below one millimeter.

Equally impressive is the mass of the universe. The Earth's mass is about 6×10^{27} g. The Sun's mass, 2×10^{33} g, is 300 thousand times greater. The mass of the Galaxy equals about 2×10^{11} solar masses. The observable part of the universe has a total mass crudely estimated as 10^{55} solar masses.

These scales cannot leave a man who tries to perceive and conceive these enormous scales and masses revealed by modern telescopes indifferent. These figures stagger the imagination and astronomers probably struggle constantly with dizziness. The immediate, natural reaction to this shock is aversion to the theory of the expanding universe. Can it be true that all the magnificence and enormousness of the universe were formerly stuck in a sphere of a few centimeters radius? And still more absurd is the question: is it possible that everything existing, everything observed could be formed "from nothing"?

Here we consciously restrict ourselves to a narrow statement of the problem. We only discuss, whether or not the assumption of the formation of the universe "from nothing" contradicts any of the firmly established general laws of nature. The point is that the most general "conservation law" is formulated as "from nothing, one can get only nothing". This "law" is naive and unscientific. The conservation laws are always concrete: e.g., the energy is conserved, or the electric charge is conserved. We will check whether or not these concrete physical laws are satisfied in the casual universe and discuss the existence of similar, more or less firmly established, physical conservation laws.

Imagine the following situation: you are visited by an inventor of some very remarkable and unusual engine or electric generator. The first reasonable step of an expert is to check whether or not you have encountered a new design of *perpetuum mobile*. It has become a custom to reject such designs without detailed analysis. The *perpetuum mobile* violates the energy conservation law, and therefore there is an inevitable mistake in the arguments of its inventor. The search for a particular mistake in a given scheme is of interest only to the inventor himself.

Let us apply this approach to the discussion of the formation of the universe "from nothing". Does this hypothesis contradict the laws of physics? Is it possible, now or in the future, to formulate a consistent, correct theory of this grandiose event?

10.2. CONSERVATION OF CHARGES

Let us begin with the conservation law for the electric charge. With this law, the answer is simple and obvious: this law does not prohibit the formation of an electrically neutral universe, i.e., the universe containing equal numbers of positive and negative charges. There exists strong evidence that this is what our universe is. Otherwise, strong electric fields would arise that would destroy the homogeneity and isotropy of the universe. Thus, the universe is probably exactly neutral and, therefore, could be formed "from nothing" without violating the conservation of electric charge.

Consider now the conservation of the baryonic charge. Recall that all known processes occurring in the laboratories conserve the total number of protons and neutrons. In particular, nuclear radioactivity is manifested either by the regrouping of protons and neutrons or by the transformation of protons into neutrons and vice versa.

When γ-rays (i.e., photons) are emitted, such regrouping occurs with the transition of the nucleus from a state with increased energy to the basic state or a state with a lower excitation energy. In α-decay, most of the protons and neutrons of the parent nucleus remain in the nucleus while some leave the nucleus combined into He nuclei (two protons plus two neutrons). During a β-decay, a fast electron (β-particle) and neutrino are produced in the transformation of a neutron into a proton. The reverse process also occurs when positrons are produced in the transformation of a proton into a neutron ($p \rightarrow N + e^+ + \nu_e$), but this process is possible only when the proton is within a nucleus and the resulting neutron occupies a lower energy state.

A free proton has a smaller mass than a free neutron, therefore, the free neutron is β-radioactive; a free proton is stable and its instability develops only within some nuclei.

This is how the conservation law of the baryonic charge was formulated at the end of the 1940's: the total number of protons and neutrons is invariant. This was then followed by the discovery of the so-called strange particles. Initially, they were discovered in cosmic rays and later thoroughly studied in accelerators. These particles are unstable. They are produced from protons and neutrons and decay into protons and neutrons. For example, $p + N = \rightarrow \Lambda + K^+ + N$ (Λ is the strange hyperon and K is the strange meson). These particles are called strange because notwithstanding the comparatively large probability of their production during a very short collision time, they have a rather long life-time of 10^{-8} to 10^{-10} s.

In the early 1950's, the so-called baryon resonances were discovered. The energy dependence of the scattering of π-mesons on protons and neutrons indicates that the latter two particles first combine into a single particle which decays a little bit later. Thus, for instance,

$$p + \pi^+ \rightarrow \Delta^{++} \rightarrow p + \pi^+,$$

$$p + \pi^- \rightarrow \Delta^0 \begin{array}{l} \rightarrow p + \pi^-, \\ \rightarrow N + \pi^0. \end{array}$$

After the discovery of these processes the baryon conservation law has acquired a more complicated form: what is conserved is the sum $B = p + N + \Lambda + \Sigma + \ldots + \Delta^{++} + \Delta^+ + \Delta^0 + \Delta^- + \ldots =$ const., or the total number of baryons.

In 1955, the long-expected experimental discovery of antiprotons occurred. The existence of antiparticles (antibaryons) was predicted shortly after the prediction and detection of antielectrons, i.e., positrons. However, the energy required for production of the pair $p + \bar{p}$ is 2000 times the energy required for the pair $e^- + e^+$. This explains why these two detections were separated by a quarter century. At that time the nerves of some scientists cracked and the very existence of antibaryons was argued against. Now there is no place for such doubts!

Thus, the final form of the baryonic charge conservation law reads $B = p + N + \Lambda + \ldots + \Delta^{++} + \ldots - \bar{p} - \bar{N} - \bar{\Lambda} - \ldots - \bar{\Lambda}^{++} - \ldots =$ const., i.e, the difference between the total numbers of baryons and antibaryons is invariant.

In the past 20 years, it was shown that a baryon consists of three quarks. The antibaryons consist of antiquarks. Correspondingly, the baryonic charge and its conservation law can be formulated in terms of quarks as

$$3B = \sum q_i - \sum \bar{q}_k = \text{const.},$$

where q_i is the number of quarks of the i-th type; $\bar{q}_k$ is the number of antiquarks of the k-th type; the sum is over all types of quarks.

The law of conservation of baryonic charge is of paramount importance both for the universe as a whole and for our immediate surroundings. This law dictates that a given number of baryons can be used for energy production only by means of their transformation into the lowest-energy state, namely, the iron nuclei whose binding energy is maximal. Thus, the energy can be produced either by transformation of uranium into the elements from the middle of Mendeleev's table or by transformation of hydrogen into iron.

The former process is successfully realized at nuclear power plants. The latter proceeds within stars. In a modified form (beginning not from the start and ending before the end), the latter process is realized in the fusion of the nuclei of deuterium and tritium with the production of ^{4}He and a neutron. In the future, this process will be the source of thermonuclear energy on Earth. The common property of both these processes is the release of only a small fraction, i.e., less than 1 percent, of the total energy stored in the fuel.

According to Einstein's equivalence law, $E = Mc^2$, the total energy stored in 1 g of matter is 9×10^{13} joules.

The abolition of the baryonic charge conservation law would imply the principal possibility of direct decay of the proton: $p = e^+ +$ energy or $p + e^- =$ energy.

Then protons, either free or bounded in nuclei, could be unstable and decay releasing enormous energy, provided that the baryonic charge conservation law was not true. The utter importance of this conservation law is obvious.

The baryonic charge of "nothing" is evidently zero. Since baryonic charge is conserved, the universe formed "from nothing" must have vanishing net baryonic charge, i.e., equal amounts of matter and antimatter. This is what was thought in the 1960's by those who first proposed the idea of the birth of the universe. They believed that the universe was born with equal numbers of baryons and antibaryons. However, if matter and antimatter in equal quantities were uniformly distributed in space (i.e., their densities were equal at every position), they would be annihilated completely when the temperature became cooler. There is no mechanism for their separation because gravity acts identically on both the matter and the antimatter.

The birth of the universe in the form we are observing is possible only in the case of the baryonic charge conservation law being violated. Avoiding a discussion of the fascinating but complicated development of the problem, we outline here the results.

The electric charge must be conserved because this is dictated by the Maxwell equations. In other words, the connection between electric charges and electromagnetic field automatically results in the conservation of the electric charge.

However, there is no field which would play an analogous role in ensuring the conservation of the baryonic charge. The belief in the conservation of the baryonic charge was based solely on experiment. Any experiment necessarily has restricted accuracy. Extrapolating experimental results, physicists, before the 1960's, tacitly assumed that strongly different numbers could appear in the world of elementary particles.

When a neutron decays into a proton (β-decay), the average decay time is about 1000 s. It would seem that nature must choose between two extremes: either a relatively fast decay of the type of the β-decay of neutrons, or no decay at all, as in the case of the absolutely stable electron. The third intermediate possibility of slow decay seemed inelegant and very improbable until the mid-sixties.

Later, tastes changed and the number of people who base their action on the presumption that everything that is not forbidden is allowed grew. Now it is admitted that the proton also can possibly decay. However, the situation remains dramatic: experimental results have shifted the lower limit of the proton decay time to 10^{32} years and the decay still has not been detected.

The belief in the non-conservation of baryonic charge is now based on the fact that the universe contains matter and is practically free of antimatter. These arguments also invoke differences in the properties of particles and antiparticles and violation of thermodynamic equilibrium due to the expansion of the universe (this was first noted by A.D. Sakharov in 1967). The corresponding estimates show that the total number of protons and neutrons is a billion times smaller than the number of photons and neutrinos. But what is most important is the understanding that emerged of the difference between the electric and the baryonic charges. In addition, the physics community (or, at least, the theoretical physicists) have got rid of the fear of large numbers. If the lifetime of the proton is as long as 10^{40} years, this would probably make proton decay detection impossible in direct experiments for many years to come. At present, it is unclear which indirect experiments would help to solve this problem.

This situation is highly appreciated by astronomers because these are astronomical observations which would lead to further progress in physics, as did the measurements of the speed of light and the verification of Newton's law of gravitation.

The existence of a universe filled with matter is as yet simple, but it is an important proof of the non-conservation of the baryonic charge!

10.3. CONSERVATION OF ENERGY

Consider now the conservation of energy in the universe as a whole. Recall that the energy of a particle at rest is equivalent to its mass, $E = Mc^2$. The conservation of the rest energy is essentially the conservation of the mass.

J. Dalton and W. Praut were the ones who noticed that the atomic weights of many elements are integers. This observation has naturally led to the hypothesis that all nuclei are formed from identical elementary bricks. The fact that the electric charges of the nuclei are not proportional to their weight implies that there exist two modifications of such bricks, protons and neutrons, which differ in their charges but have nearly equal masses. Here we somewhat deviate from the description of the historical development, having omitted the dismal period when the nuclei were believed to be formed from protons and electrons. Roughly speaking, due to the uncertainty principle, electrons cannot fit into the small nuclei. The first correct ideas of the existence of neutrons were proposed as a hypothesis in the early 1920's. Rigorous proofs of their existence came in the 1930's and in 1945 the Hiroshima and Nagasaki bombings occurred. In this very brief account of the historical development, we have not mentioned the discovery of isotopes and the very accurate determination of atomic weights of individual isotopes.

As a result, on the one hand, the theory of the universal composition of nuclei by protons and neutrons has been substantiated; on the other hand, the first argument in favour of this theory — the integer atomic weights of isotopes — turned out to be inaccurate. This is the irony of the development of science. However, the not exactly integer atomic weights of isotopes have acquired another, no less important, significance.

From the fact that the mass of helium atom is 0.6 percent smaller than the combined mass of four atoms of hydrogen, astrophysicists have deduced that in the interiors of stars, hydrogen transforms into helium releasing 0.6 percent of its mass ($0.006 \times c^2 = 5.4 \times 10^{18}$ erg g^{-1}) in the form of radiation. We should specially emphasize that this conclusion was reached long before the development of nuclear physics suggested specific ways of this transformation.

We have made this excursion to nuclear physics in order to elucidate that, somehow, released gravitational energy also results in a decrease of mass of the whole in comparison with the sum of masses of the parts. The mass of a neutron star is around 10–15 percent smaller than the total mass of its particles. This difference of masses fuels the supernova explosions that accompany the formation of neutron stars, even though a very large fraction of this energy is carried away by the neutrinos.

Is the mass fraction that can be converted into energy by gravitation limited? As early as 1962, one of the authors showed that such limit does not exist (Zeldovich, 1962). Massive bodies, whose mass exceed 2–3 solar masses, reach the high-density state in the course of their natural evolution. Low-mass bodies can reach high densities only after overcoming a very high barrier. A weight of iron of mass 1 kg is practically stable; it is curious that having spent the energy E to compress it, one can release the energy $E + 999\,\mathrm{g}\,c^2$ in the course of subsequent spontaneous collapse, i.e., transform 999 grams of mass into energy. After the collapse, the mass of the weight would decrease to 1 g and its final size would be inconceivably small, of the order 10^{-28} cm. This small fact is associated with very sad memories: it was the last piece of discussion between one of the authors (Ya.B.Z.) and his teacher, L.D. Landau, a few days before the tragic accident.

In their remarkable book on field theory, L.D. Landau and E.M. Lifshitz gave an exact and rigorous proof of the fact that the mass (and, consequently, the energy) of a closed world is identically zero. The discussion above allows us to understand this statement intuitively. The negative energy of the gravitational interaction between the parts exactly compensates for the positive energy of the whole of all the matter present. The general relativity theory establishes a connection between gravitation and geometry. This proves that exact compensation occurs if the space that contains matter is closed.

Thus, the general relativity theory removes the last obstacle for the formation of the universe "from nothing". The energy of "nothing" is zero. But the energy of a closed universe is also zero. Therefore, the energy conservation law does not preclude the formation of a closed universe "from nothing" (it is important that what is allowed is a geometrically closed universe but not an open infinite universe).

10.4. THE ASTROPHYSICAL CONSEQUENCES. IS THE PULSATING UNIVERSE INEVITABLE?

The first consequence is that the total density of all forms of matter must be sufficiently high; this lends additional support to the existence, in some form, of the "hidden mass" because the density of the ordinary and well known forms of mass (protons, nuclei, electrons, photons) is insufficient.

The second consequence is that the expansion of the universe as observed now should be replaced in future by contraction. This contraction will occur earlier or, more probably, later even in comparison with the present age of the universe.[a]

The idea of a closed world, which first expands and then contracts, has

[a] However, note another variant of the cosmological theory (see Zeldovich, 1984): there is a possibility that the cosmological constant is non-vanishing and has such a sign that it replaces a part of the mass. Then the expansion proceeds indefinitely – the world is not "closed" along the time axis (there was the time of birth but general collapse will never occur) even though it is closed spatially.

led to the hypothesis of a pulsating endless universe. One small thing is lacking in both literal and figurative senses: we understand how expansion stops and is replaced by contraction at the *large* (maximal) radius of the universe; what remains to be understood is how the transition from contraction to expansion at the *small* (minimal) radius occurs. The idea of the universe being infinite in time (to the past!) became more popular when it was realized that when allowance was made for vacuum polarization by spatial curvature (strong gravitational field) or by the gravitational field produced by the scalar field with non-zero mass, the equations of general relativity indeed possess formally rigorous inflationary solutions of the type

$$a(t) = \frac{1}{H_0} \cosh (H_0 t) = (e^{-H_0 t} + e^{+H_0 t})/2H_0,$$

for the radius of the universe, with the minimal radius H_0^{-1} of the order 10^{-28} cm (here H_0 is the Hubble constant for this early epoch). Formally, these solutions are also present in the classical theory. Which objections can be brought forward against these solutions?

We consider the possibility of the formation of the universe "from nothing" as the most important objection. The idea of an endless universe seemed to be inevitable (and only the way of transition from contraction to expansion, in particular, whether it is classical or quantum, was debated) until we came to believe that the energy and the baryonic charge were forever given as conserved and non-vanishing quantities. Now, we have got rid of the charms of these ideas. Having agreed that the hypothesis of the endless universe is not inevitable, we can turn to a discussion of the details of the theory of cyclic evolution.

As early as in the 1930's, a serious thermodynamic argument was put forward against the idea of an everlasting, cyclically reproduced universe. Entropy grows during every cycle.[b] As a result, the amplitude of pulsations grew from one cycle to the subsequent one. Applying this

[b] There exists the viewpoint that entropy growth is replaced by its decrease when expansion is replaced by contraction. In this connection, a mysterious reversal of the "time arrow" is mentioned. We note that the influence of the overall slow expansion or contraction on specific processes involving elementary particles or stellar interiors seems to be unphysical and groundless.

result to the past, we see that the total number of cycles, beginning with the first one with zero entropy, is finite. Thus, a cyclically evolving universe nevertheless has a finite age, i.e., it should be somehow born.

Recently Belinsky *et al.* (1985) have considered the expansion and contraction of the universe filled by a massive coherent scalar field. Similar calculations have been carried out even earlier but, perhaps, with less definite conclusions. The relation between pressure and energy density depends on whether the scalar field φ is almost steady ($m\varphi^2 \gg \hbar\dot{\varphi}^2$), or rapidly changing and not massive ($\hbar^2\dot{\varphi}^2 \gg m^2\varphi^2$). In the former case, $p = -\varepsilon$, and gravitational repulsion occurs while in the latter case the pressure is maximal, $p = +\varepsilon$, and the gravitational force is attractive.

In principle, both these situations can have a place in both contraction and expansion. However, during the contraction stage, the latter regime, $p = +\varepsilon$, is stable and the field pressure opposes contraction. In this case the classical solution predicts singularity and the radius of the universe vanishes. Solutions with a smooth transition from contraction to expansion turn out to be exceptional and improbable. However, what is more important is not the detailed analysis of the solutions but rather the analysis of the assumptions adopted in the course of the solution of the problem.

We consider an exactly homogeneous scalar field and an exactly homogeneous and isotropic universe. Homogeneity implies uniformity, i.e., equivalence of all spatial positions at any given moment. Isotropy implies the equivalence of all spatial directions.

When we consider *expansion*, these presumptions are reasonable: those regions where the scalar field is maximal rapidly expand. The classical scalar field becomes practically uniform while all the other fields (in particular, the electromagnetic field which violates isotropy) rapidly become weaker.

However, in the course of *contraction*, one can expect strong instability, violation of homogeneity and isotropy. Allowance for perturbations thus makes the possibility of smooth passing of the universe through some finite minimal radius even less probable. In essence, this argument is close to the notion of growing entropy. Hence, even if these arguments cannot be considered as a rigorous theorem, we have enough reasons to turn our minds to the spontaneous formation of the universe which removes the necessity of a cyclic universe.

10.5. THE SPONTANEOUS FORMATION

Above we have pondered over the principal possibility of the birth of the universe. What can be said about the specific mechanism of this phenomenon? Here we can only formulate the problem.

First of all, the words "nothing" and "from nothing" can be understood in various ways. One can think of empty, flat Minkowsky space — the general relativity equations have such solutions and it is not restricted in time. The birth of the universe can be envisaged as a cartoon shown in Fig. 10.1. The reader should keep in mind that what is shown is the one-dimensional analogue. We cannot sketch the birth of a three-dimensional closed space from (three-dimensional cross-section of) the Minkowsky space. The parts I–IV of the figure differ in the corresponding time t. After splitting out of a closed region, the remaining space restores its flatness. However, this space is flat only in the classical limit. Actually, the quantum theory predicts fluctuations of the space-time metric just as an oscillator has finite and equal averaged kinetic and potential energies $\overline{mv^2}/2 = \bar{u} \neq 0$ in the lowest-energy state.

Thus, we are talking about a fluctuation, but such a strong fluctuation that the very topology changes and the space splits into two parts. At present, we cannot analyze such fluctuations formally. Recall that now we can deduce the properties of a vacuum (its average energy, i.e., the cosmological constant) only from experiments.

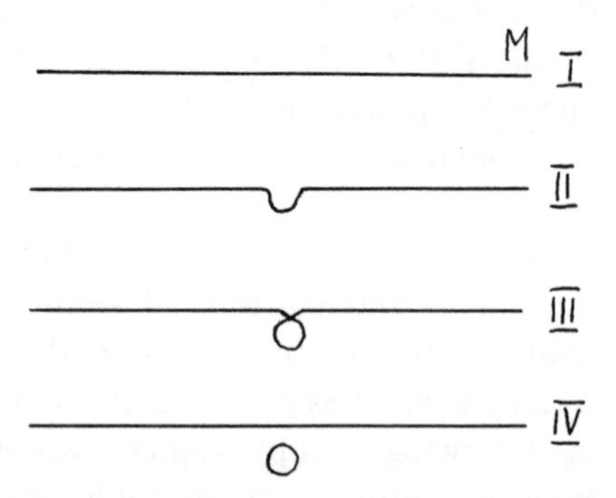

Fig. 10.1. The birth of the closed universe (a ball in the lowest panel IV) from flat Minkowsky space (M at the stage I). At intermediate stages, the metric remains flat (Minkowskian) far from the fluctuation which gives birth (splitting out) to the ball.

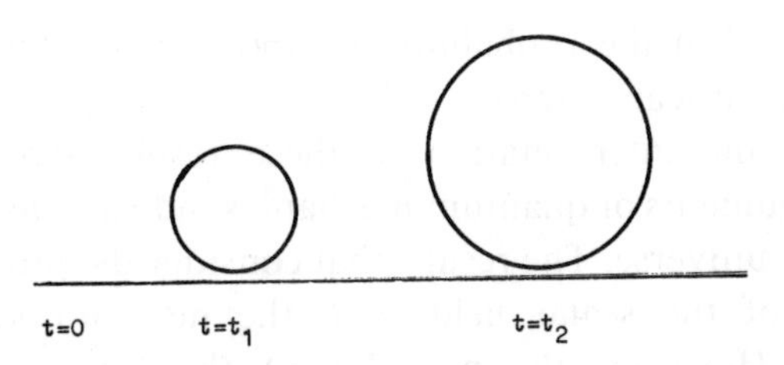

Fig. 10.2. Spontaneous formation of the universe "from nothing". Before the moment $t = 0$, the metric (and time, in particular) did not exist.

Another popular possibility considers a sole closed world (without the background of the parent Minkowsky space) (see Fig. 10.2). In this case nothing exists before the "beginning", no metric and, in particular, no time.

The classical equations of motion do not possess such relevant solutions. One should consider quantum solutions. This problem bears some similarity to the problem of α-decay of the uranium or radium nucleus. According to classical mechanics, α-particles cannot traverse all the distance from the parent nucleus to infinity. The solutions in quantum mechanics for α-particles are able to describe both regions: the "under-barrier", in which kinetic energy is negative (i.e., classical motion is impossible) and the far region where both solutions, classical and quantum, exist and differ only slightly.

The quantum-mechanical theory of the birth of the universe is formulated similarly to the theory of α-decay. Quite naturally, this problem is solved in a very crude approximation taking into account only two quantities, the radius of the closed universe $a(t)$ and the scalar field φ. In quantum theory, one introduces the corresponding momenta P_a and P_φ and constructs the wave function $\psi(a, \varphi)$. The momentum $P_a = M_{\text{eff}}\dot{a} = f(a)\dot{a}$ is proportional to the expansion rate and the classical limit allows us to evaluate $\dot{a} = da/dt$ and, hence, the time, $t = \int (da/dt)^{-1}\, da$.

Note also that the quantum theory gives, even in its present under-developed form, arguments in favour of a closed universe (in contrast to an infinite flat or open universe). Only for a closed universe can one introduce a certain finite-value effective mass M_{eff}. Whatever formulation of quantum mechanics we adopt (wave functions, or "path integration", or whatever else) the probability of spontaneous birth of an infinite universe is identical to zero.[c]

However, the interpretation of these results remains, in general, unclear. Arguments of quantum mechanics indicate the *possibility* of the origin of the universe. The results that consider the probability of this or that value of the scalar field φ in the newborn universe are also interesting. However, the modulus of the wave function and the probability are difficult to interpret when we discuss the origin of a unique object, our universe. There are also more fundamental reasons for a sceptical attitude to concrete theories about the formation of the universe "from nothing".

The point is that the development of fundamental physics is still far from its end! Moreover, recently we see a revival of theories of elementary particles relying more on geometry. On the other hand, these are the theories which unify the bosons and the fermions and also unify the internal variables of particles and fields with the coordinates and the Lorentz transformations. In perspective, these theories should give a direct proof of the existence of the scalar fields and determine their properties. Sooner or later, the theory of the masses of particles will be formulated and *physicists* will tell us what the hidden mass discovered by *astronomers* is. Even closer to the problem of the origin of the universe are hypotheses of high-dimensional space-time. The idea that there is one "excessive" coordinate x_4 coiled into a ring of length $l = 2\pi R$, where R is the ring radius, was proposed as early as in the 1920's.[d] This situation is schematically shown in Fig. 10.3 where three spatial coordinates are replaced by a single one, x, directed along the tube.

[c] Zeldovich and Starobinsky (1984) have considered a flat universe which is finite as a torus by identifying the opposite sides of a cube. However, this leads to the loss of exact isotropy of space: the directions parallel to the cube diagonals are not equivalent to those orthogonal to the sides or edges. However, this hypothesis as yet cannot be ruled out. The possibilities of observational verification of such hypotheses are discussed by Sokoloff and Shvartsman (1974).

[d] Such closure, which restricts the interval of variation of the coordinate x_4 by a small value $l(x_4 + l \equiv x_4)$, is what mathematicians call compactification. We denote this coordinate x_4, bearing in mind that time is usually denoted by x_0 and spatial coordinates x_1, x_2, x_3. The resulting space-time is five-dimensional. This theory was proposed by T. Kaluza and O. Klein in the 1920's (see Kaluza, 1921).

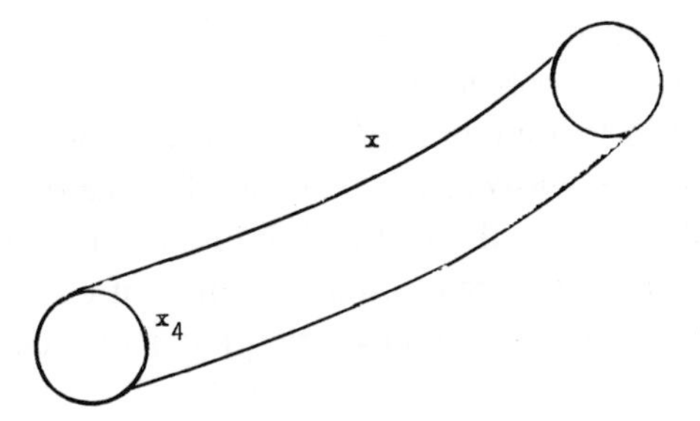

Fig. 10.3. Schematic view of the Kaluza-Klein space. Shown is the cross-section for a certain given moment, $x_0 = \text{const}$. The three spatial coordinates x_1, x_2 and x_3 are replaced in this figure by a single one, x. The resulting configuration has the form of a two-dimensional surface of a tube with coordinates x and x_4 measured along its surface.

In quantum theory, the motion of a particle along the closed coordinate x_4, or particle localization at coordinate x_4, requires the particle to have enormous energy. Thus, even in experiments involving enormous energies of 10^{17} or even 10^{19} GeV (compare with 10^3 GeV at the accelerators of the 1980's), motion along the distinguished coordinate x_4 cannot be detected (or along the coordinates from x_4 through x_9). Theorists say that in the *low-energy limit*, space-time remains effectively four-dimensional. If, in addition, we restrict ourselves to scales which are small in comparison with astronomical ones, space-time can be described by the good old metric of Minkowski.

Nevertheless, the introduction of additional dimensions, x_4 in the simplest example, does not occur without any consequence. One can envisage weak variations of the metric for which the additional coordinate axis is presumed to be non-orthogonal to the basic (macroscopic) axes. This presumption turns out to be equivalent to introduction of the electromagnetic field in ordinary space.

The introduction of more compactified ("closed") variables of up to 6 or 7 (transition to 10-dimensional space-time) allows us to introduce not only the electromagnetic field but also those fields ($W^{\pm}$, Z_0) which describe weak interaction as well as gluon fields responsible for strong interaction. In addition, the theory of supersymmetry which unifies boson fields (in particular, the electromagnetic field) and fermion fields (in particular, the electron-positron field), is also "geometrical". It introduces unusual variables but they are geometrical. Einstein's dream of geometrization of the whole of physics is much more realistic today than it seemed five or ten years before.

We at last show a formulation of theories in which the scalar fields become the inevitable *consequence*. The significance of the scalar fields was mentioned above — without them, the inflationary universe would be impossible. Vacuum polarization as a source of energy and negative pressure for inflation of the universe are a particular case of the scalar field.

However, what is more important and specific is the other aspect of the influence on cosmology of the theories with "excessive" dimensions. At the moment of birth of the closed world, the spatial variables x_1, x_2 and x_3 vary within a very narrow range of the order of $2\pi a(t)$, where $a(t) \rightarrow 0$ for $t > 0$. It is natural then to presume in this case that the universe is actually born symmetric over all the spatial variables (of dimension 5 or more). The separation of geometrical variables into "internal", or compactified, variables and the usual three geometric variables occurs only later. This separation is actually a typical spontaneous violation of symmetry! Initially we have, for instance, a 9-dimensional sphere in which all directions are equivalent, and later six dimensions from a_4 through a_9 become frozen (their characteristic scales remain of the order 10^{-33} cm) while the remaining three dimensions grow exponentially and eventually exceed 5000 Mpc = 10^{28} cm, i.e., they become larger than the observable part of the universe.

Our intention remains the same as before. We wish to describe the formation of the universe "from nothing". However, the concrete realization of this intention becomes drastically different from the initial variants.

Thus, further progress in cosmology requires a fundamental development of the physics of the micro-world. Not only the *grand unification* of various interactions but also the *super grand unification* of the micro-world and cosmology present the most fundamental and ambitious programme of the final decades of the twentieth century.

EPILOGUE

THE BIRTH OF "DIVINAMICS"

Until recently, chance was not very influential in physical theories. Theorists prefered deterministic theories. Such classical giants as mechanics, relativity theory and electrodynamics do not deal with randomness. The laws of chance were invoked only for the description of enormous particle ensembles in statistical physics.

Randomness is an integral part of quantum mechanics but the Schrödinger equation for the basic quantity or the wave function is strictly deterministic. Only the evaluation of the observable quantities relies on probability theory.

In fact, the generality of the deterministic approach is only an illusion. Only in special cases are solutions totally deterministic and predictable. Actually, even the trajectories of relatively simple mechanical systems cannot be predicted.

Of course, if Laplace's demon could determine exactly the coordinates and velocities of all the particles at some initial moment, the future would become absolutely predictable. However, when two infinitesimally close trajectories diverge exponentially, any uncertainty in the initial conditions rapidly grows and the prediction for sufficiently long periods becomes impossible. It is clear that the initial conditions cannot be determined exactly because experimental errors are unavoidable. But it is not always appreciated (at least, it was not realized during the times of Laplace) that this cannot be done even in principle. The problem is associated with the difficulties of formulating the very concept of very large and very small numbers.

In order to define a number, for instance, through n figures of a binary sequence, n bits of information are required. Only some numbers, e.g., periodic fractions, can be described with less information. Therefore, using only the rules (algorithms) of finite complexity, one can represent only a relatively small set of numbers. In particular, the exact evaluation of almost every real number requires an algorithm of infinite length. It is interesting that the number π can be evaluated by a infinite recursion of a short algorithm (see, e.g., Brudno, 1978). As M. Kac wittily mentioned, almost all numbers from the continuum cannot be defined through a finite number of words (see the paper of Ford (1983) devoted to these problems). Moreover, some mathematicians (see Rashevsky, 1976) pose the problem of whether or not the notion of a very large integer number is consistent. The point is that, when counting is understood as the result of sorting of similar obejcts, then sooner or later there arises the problem of a more exact definition of the given class of objects (see the remarkable paper of Essenin-Volpin (1966)). For instance, when the reader counts the cars passing by, he will eventually have the problem of whether or not a three-wheel vehicle should also be counted.

Thus, randomness is inevitable in the initial conditions. The exponential instability, typical of mechanical and other systems, makes the role of chance essential after a sufficient time. We should reject hopes of deterministic prediction for such long periods. It does not imply, however, that we are left to the mercy of blind fate. Even in casual situations, there are elements of predictability which admit important, if not complete, knowledge. One such element is intermittency, i.e., structures in random media. One can predict the characteristic size, lifetime, amplitude and some other properties of these structures. To some extent, they behave as regular objects. Note, however, that their exact positions and moments of appearance cannot be known in advance. Of course, the intrinsic determinacy of a random process has deep roots in the central limit theorem.

The inevitability of randomness and combinations of random and deterministic properties are thus a very general aspect of the physical universe. In ancient Rome, the prediction and promotion of such events which were believed to be occasional were called "divinatio" (see Cicero's treatise 1964, *De divinatione*). In appreciation of the deepness of the insight of the ancient philosophers, the dynamics that takes inevitable randomness into account can be called *div*inamics.

REFERENCES

V. S. Afraimovich, V. V. Bykov and L. P. Shilnikov, "On the origin and structure of the Lorenz attractor", *Dok. Akad. Nauk.* (SSSR) **234** (1977) 336–339.

M. E. Agishtein and A. A. Migdal, "Computer simulation of three-dimensional vortex dynamics", Preprint Space Research Institute No. 1102 (1986).

A. P. Aldushin, Ya. B. Zeldovich and B. A. Malomed, "On phenomenological theory of spin combustion", *Combust. and Flame* **42** (1981) 1–6.

H. Alfven, "Discussion of the origin of the terrestrial and solar magnetic field", *Tellus* **2** (1950) 74–79.

A. A. Andronov, "Les cycles limités de Poincaré et la théorie des oscillations auto-entretenues", *C. R. Acad.Sci.* (Paris) **189** (1929) 559–561.

F. Anselment, Y. Gagne, E. J. Hopfinger and R. A. Antonia, "High-order velocity structure functions in turbulent shear flows", *J. Fluid Mech.* **140** (1970) 63–90.

V. I. Arnold, *Mathematical Methods of Classical Mechanics*, Springer-Verlag (1980).

V. I. Arnold, "Some comments on anti-dynamo theorems", Vestn. Moscow Univ., ser. I, *mat. mekh.* No. 6 (1982) 50–57.

V. I. Arnold, "Singularities, Bifurcations and Catastrophes", *Usp. Fiz. Nauk* (SSSR) **141** (1983) 569–590.

V. I. Arnold, Ya. B. Zeldovich, A. A. Ruzmaikin and D. D. Sokoloff, "Magnetic field in a stationary flow with stretching in a Riemannian space", *JETP* **81** (1981) 2052–2058; *Sov. Phys. JETP* **56** (1981) 1083–1086.

O. V. Artamonova and D. D. Sokoloff, "Asymptotic analysis of the fast growth of the second moment of magnetic field in mirror-symmetric flow". Vestn. Moscow Univ., ser. II, *fiz. astron.* **27** (1986) No. 38–13.

V. M. Babich and V. S. Baldyrev, *Asymptotic Methods in Problems of Short Waves Diffraction*, Nauka, Moscow (1972).

L. Bachelier, "Théorie de la spéculation", *Ann. Sci. Ecole Norm. Sup.* **17** (1900) 21–86.

J. Yu. Bakelmann, A. L. Werner, B. E. Kantor, *An Introduction to Differential Geometry as a Whole*, Nauka, Moscow (1973).

N. B. Baranova and Ya. B. Zeldovich, "Dislocations of wave front surfaces and amplitude zeros", *JETP* **80** (1981) 1789–1797.

N. B. Baranova, Ya. B. Zeldovich, A. V. Mamayev, N. P. Pilipetsky and V. V. Shkunov, "Dislocation density of the wave front with a speckle structure of the light fields", *JETP* **83** (1982) 1702–1710.

G. I. Barenblatt, *Similarity, Self-Similarity and Intermediate Asymptotics*, Plenum, New York (1980).

G. I. Barenblatt and Ya. B. Zeldovich, "Self-similar solutions as intermediate asymptotics". *Usp. Mat. Nauk* **26** No. 2 (1971) 115–129; *Russ. Math. Surveys* **26** No. 2 (1972) 45.

G. K. Batchelor, "On the spontaneous magnetic field in a conducting fluid in turbulent motion", *Proc. R. Soc. London* **A201** (1950) 405–416.

V. A. Belinsky, L. P. Grishchuk, I. M. Khalatnikov and Ya. B. Zeldovich, "Inflationary stages in cosmological models with scalar fields", *JETP* **89** (1985) 346–355.

W. A. Berggren and J. A. van Couvering (eds.), *Catastrophes and Earth History. The new Uniformitarianism*, Princeton Univ. Press (1984).

M. V. Berry, J. F. Nye, "Dislocations in wave trains", *Proc. R. Soc. London* **A336** (1974) 165–190.

E. Borel, "Sur les probabilités denombrables et leurs applications arithmétiques". *Rend. Cioc. Mat. Palermo* **47** (1909) 247–271.

C. Brooks, S. R. Hart and J. Wendt, "Realistic use of two-error regression treatment as applied to rubidium-strontium data", *Rev. Geophys. and Space Phys.* **10** (1972) 551–577.

R. Brown, "A brief report on microscopic observations, conducted in June, July and August 1827, of particles in pollen of plants", *Philos. Mag. Ann of Philos.* (New Ser.) **4** (1828) 161–178.

A. A. Brudno, "On complexity of trajectories of a dynamic system" *Usp. Mat. Nauk* **33** No. 1 (1978) 207–208.

J. Brünning, "Über Knoten von Eigenfunktionen des Laplace-Beltrami-Operators", *Mathem. Z.* **158** (1978) 15–21.

L. A. Bunimovich, "On the decrease of correlations in dynamic systems with chaotic behaviour", *JETP* **89** (1985) 1463–1482.

L. A. Bunimovich and Ya. G. Sinai, "Statistical properties of Lorenz gas with periodic configurators of scatterers", *Commun. Math. Phys.* **78** (1981) 479–497.

P. Cartier, "Perturbations singulières des équations différentielles ordinaires et analyse non-standard", Séminaire Bourbaki, 1981/82, Soc. Math. France, Paris (1982) 21–44.

M. T. Cicero, *De senectute*, *De amicite*, *De divinatione*, With English transl. W. A. Falconer, London, Cambridge (Mass.): Loeb Classical Library, 1964.

R. Courant and D. Hilbert, *Methods of Mathematical Physics*, Springer-Verlag (1962).

R. D. Dakgesamansky and V. R. Shutenkov, "Variation of the background radiation and structure of galactic magnetic field", *Pisma Astron. J.* (SSSR) **13** (1987) 182–190.

R. G. Deissler, "Turbulent solutions of the equations of fluid motion", *Rev. Mod. Phys.* **56** (1984) 223–254.

P. Dittrich, S. A. Molchanov, A. A. Ruzmaikin and D. D. Sokoloff, "The limiting distribution of magnetic field strength in a random flow", *Magnitnaya gidrodinamika* No. 3 (1988) 9–12.

A. Dold and B. Eckmann, (eds.), *Stability Problems for Stochastic Models*, Springer-Verlag (1984).

I. M. Dryomin, "Fractals in problems of multiple particle production", *Pisma JETP* **45** (1987) 505–507.

A. Dewdney, "A computer microscope zooms in for a loop at the most complex object in mathematics", *Sci. Am.* **253** No. 8 (1985) 16–24.

T. Dombre, U. Frisch, J. M. Greene, M. Hénon, A. Mehr and A. M. Soward, "Chaotic steamlines and Lagrangian turbulence: the ABC flows" *J. Fluid Mech.* **167** (1986) 353–391.

G. M. Dvorina, A. A. Ruzmaikin and D. D. Sokoloff, "Temperature spots in turbulent flow with heating," *Teplofizika vysokykh temperatur* (SSSR), in press (1989).

J. P. Eckmann and D. Ruelle, "Egrodic theory of chaos and strange attractors", *Rev. Mod. Phys.* **57** (1985) 617–656.

A. L. Efors, *Physica and Geometry of Disorder*, Nauka, Moscow (1982).

A. Einstein, *Eine neue Bestimmung der Moleküldimensionen*, Buchdruck K. J. Wyess, Bern (1905).

A. Einstein, "Zur Theorie der Brownschen Bewegung", *Ann. Phys.* **19** (1906) 371–381.

A. S. Essenin-Volpin, Editor's notes in Russian translation of *Foundations of Set Theory* by A. A. Fraenkel and J. Bar-Hillel, Mir Publ., Moscow (1967).

D. Farmer, "Chaotic attractors of an infinite-dimensional dynamical system", *Physica D* **4** (1982) 366–393.

W. Feller, *An Introduction to Probability Theory and its Applications*, vols. 1–2, John Wiley & Sons (1966).

H. Ferstenberg, "Noncommuting random products", *Trans. Am. Math. Soc.* **108** (1963) 377–428.

R. P. Feynman and A. R. Hibbs, *Quantum Mechanics and Path Integrals*,

McGraw-Hill (1965).

J. Finn and E. Ott, "Chaotic Flows and Fast Magnetic Dynamos", *Phys. Rev. Lett.* **60** (1988) 760–763.

J. Ford, "Is the outcome of coin tossing random?", *Physics Today*, No. 3 (1983) 40–46.

A. A. Fraenkel and V. Bar-Hillel, *Foundations of Set Theory*, North-Holland, Amsterdam (1958).

U. Frisch and G. Parisi, "On the singularity structure of fully developed turbulence" in *Turbulence and Predictability in Geophysical Fluid Dynamics and Climate Dynamics*, LXXXVIII Corso, Soc. Italiana di Fisica, Bologna, Italy (1985) 84–88.

A. V. Gaponov-Grehkov and M. I. Rabinovich, "L. I. Mandelshtam and the modern theory of non-linear oscillations and waves", *Usp. Fiz. Nauk* (SSSR) **128** (1979) 579–624; *Sov. Phys. Usp.* **22** (1979) 590–614.

C. W. Gardiner, *Handbook of Stochastic Methods for Physics, Chemistry and Natural Sciences*, Springer Series in Synergetics, **13**, H. Haken (ed.), Springer-Verlag (1980).

I. M. Gelfand and A. M. Yaglom, "Functional-space integration and its applications to quantum physics", *Usp. Mat. Nauk* **11** (1956) 77–114.

I. I. Gikhman and A. V. Skorokhod, *Theory of Random Processes*, vols. 1–3, Nauka, Moscow (1971, 1973, 1975).

K. G. Giovanelli, *Secrets of the Sun*, Cambridge Univ. Press (1984).

E. B. Gledser, F. V. Dolzhansky, and A. M. Obukhov, *The System of Hydrodynamic Type and their Applications*, Nauka, Moscow (1981).

E. S. Golubtsova and T. A. Koshelenko, "Against distortions of ancient history", *Voprosy istorii* (Problems of history) No. 8 (1982) 70–82.

P. Grassberger and I. Procaccia, "Measuring the strangeness of strange attractors" *Physica D* **9** (1983) 189–208.

Gregorii episcopi Turonesis, *Historiarum libri decem*, ed. H. Büchner, vols. 1–2 Berlin (1956).

M. L. Gromov, "Isometric imbeddings and immersions", *Dok. Akad. Nauk* (SSSR) **192** (1970) 1206–1209; *Sov. Math. Dokl.* **11** (1971) 794.

J. J. Halliwell and S. W. Hawking, "Origin of structure in the Universe", *Phys. Rev.* **D31** (1985) 1777–1791.

G. H. Hardy, "Weierstrass's non-differentiable function", *Trans. Am. Math. Soc.* **17** (1916) 301–325.

K. Hasselmann, "Stochastic climate models. 1. Theory", *Tellus* **28** (1976) 473–485.

F. Hausdorff, "Dimension and äusseres Mass", **79** (1918) 157–179.

M. M. Hénon, "Sur la topologie des lignes de courant dans un cas particulier",

C. R. Acad. Sci. (Paris) **262** (1966) 312–314.
C. Hermite, *Correspondance d'Hermite et de Stieltjes*, **2**, p. 318, ved. S. Bailland and H. Bourget, Gauthier-Villars, Paris (1905).
T. Hida, *Brownian Motion*, Springer-Verlag (1980).
D. Hilbert, "Über das Unendliche", *Math. Annalen* **95** (1925) 161–190.
E. Hopf, "Statistical hydromechanics and functional calculus", *J. Rat. Mech. Anal.* **1** (1952) 87–123.
W. von Humboldt, "Über den Nationalcharakter der Sprachen", *Gesammelte Schriften*, **4**, Berlin (1905).
A. K. M. F. Hussain, "Coherent structures and turbulence", *J. Fluid Mech.* **173** (1986) 303–356.
"Information concerning the accident at the Chernobyl nuclear power station and its consequences prepared for IAEA", *Atomnaya Energiya* (Atomic Energy, SSSR) **61** (1986) 301–320.
K. Ito, "On a stochastic differential equation", *Mem. Amer. Math. Soc.* **4** (1951) 1–5.
G. R. Ivanitsky, V. I. Krinsky and E. E. Selkov, *Mathematical Biophysics of Cells*, Nauka, Moscow (1978).
K. Janicka, *Surrealizm*, Wydanictwa Artystyczne i Filmowe, Warszawa (1985).
T. W. Jones, "Polarization as a probe of magnetic field and plasma properties of compact radio sources: simulation of relativistic jets", *Ap. J.* **332** (1988) 678–695.
M. Kac, *Probability and related topics in physical sciences*, Interscience (1957).
B. B. Kadomtsev and O. P. Pogutse, "Electron heat conductivity of the plasma across a 'braided' magnetic field", in *Plasma Physics and Controlled Nuclear Fusion Research*, Proc, VII Intern. Conf., Innsbruck, 1978, IAEA, Vienna (1979) 649–663.
A. P. Kazantsev, "Enhancement of a magnetic field by a conducting fluid", *JETP* **53** (1967) 1806–1813; *Sov. Phys. JETP* **26** (1968) 1031–1039.
H. Kesten, *Percolation theory for Mathematician*, Birkhäuser (1962).
P. L. Kirillov, "On the influence of thermodynamical properties of a surface on heat fluxes in turbulent flow", *Inzh. fiz. zhurn.* (SSSR) No. 3 (1966) 501–512.
I. M. Kirko, F. M. Mitenkov, V. A. Barannikov, A. K. Gailitis, A. S. Karabasov, G. E. Kirko, A. I. Kiriushin, O. A. Lielausis, A. M. Mukhametshin, V. N. Ponomaryov, M. T. Telichko and A. P. Sheykman, "Observation of MHD phenomenon within the volume of liquid metal of the cooling system of the breeder reactor BN-600 at Beloyarsk Nuclear Energy Plant", *Dok. Akad. Nauk* (SSSR) **257** (1981) 861–863.
N. L. Kleeorin, A. A. Ruzmaikin and D. D. Sokoloff, "Correlation properties of self-exciting fluctuative magnetic fields" in *Plasma Astrophysics*, ESA SP-251,

Paris (1986) 539–544.
A. N. Kolmogoroff, *Basic Principles of Probability Theory*, Gostekhizdat, Moscow (1936).
A. N. Kolmogoroff, "Zur Umkehrbarkeit der statistischen Naturgesetze", *Math. Annalen* **113** (1937) 766–772.
A. N. Kolmogoroff, *Probability Theory and Mathematical Statistics*, Nauka, Moscow (1986).
A. N. Kolmogoroff, I. G. Petrovsky and N. S. Piskunov, "The analysis of equation of diffusion accompanied by growth of quantity of matter and its application to certain biological problem", *Bull. Moscow State Univ.*, Sect. A **1** No. 6 (1937) 1–26.
R. H. Kraichnan, "Diffusion of passive scalar and magnetic field by helical turbulence", *J. Fluid Mech.* **77** (1976) 753–768.
R. H. Kraichnan and S. Nagarajan, "Growth of turbulent magnetic fields", *Phys. Fluids* **10** (1967) 859–870.
F. Krause and K. H. Rädler, *Meanfield Magnetohydrodynamics and Dynamo Theory*, Pergamon Press (1980).
T. Kaluza, "Zum Unitätsproblem der Physik", *Sitz. Press Akad. Wiss., S.* (1921) 966–980.
L. D. Landau, "On the problem of turbulence", *Dok. Akad. Nauk* (SSSR) **44** (1944) 339–341.
L. D. Landau and E. M. Lifshitz, *Fluid Mechanics*, Addison-Wesley, Reading, Mass. (1959).
L. D. Landau and E. M. Lifshitz, *Statistical Physics*, Nauka, Moscow (1964).
L. D. Landau and E. M. Lifshitz, *The Classical Theory of Fields*, Addison-Wesley, Reading, Mass. (1962).
V. P. Lazutkin, "Estimation of the gap width in spectrum of the Laplace operator for a flat convex region", *Dok. Akad. Nauk* **245** (1979) 20–23.
J. Larmor, "How could a rotating body such as the Sun become magnetic?", *Rep. Brit. Assoc. Adv. Sci.* (1919) 159–160.
P. Levy, *Processes stochastiques et mouvement Brownien*, IInd ed., Paris (1965).
E. H. Lorenz, "Deterministic nonperiodic flow", *J. Atomos. Sci.* **20** (1963) 130–141.
M. Lücke, "Statistical dynamics of the Lorenz model". *J. Stat. Phys.* **15** (1976) 455–475.
I. M. Lifshitz, S. A. Gredeskul and L. A. Pastur, *Introduction to the Theory of Disordered Media*, Nauka, Moscow (1982).
N. N. Luchinsky, "Energy consumption estimates for a vehicle in an unsteady motion", *Dokl. VASKHNIL* (CR. Sov. Akad. Agricult. Sci.) No. 11 (1984) 41–43.

A. F. Maistrov, *Development of the Concept of Probability*, Nauka, Moscow (1980).
M. Maksimov, "Are 'random' numbers random?", *Nauka i Zhizn* (Science and Life, USSR) No. 10 (1986) 112–113.
B. Malraison, P. Atten, P. Berge, and M. Dubois, "Dimension of strange attractors: an experimental determination for the chaotic regime of two convective systems", *J. Phys. Lett.* **44** (1983) L897.
B. Mandelbrot, "Possible refinement of the lognormal hypothesis concerning the distribution of energy dissipation in intermittent turbulence", *Lect. Not. Phys.* **12** (1972) 333–351.
B. Mandelbrot, "Fractals and turbulence: attractors and dispersion", *Lect. Not. Math.* **615** (1977) 83–93.
B. Mandelbrot, *Fractals: Form, Chance and Dimension*, Freeman, San Francisco (1977).
V. P. Maslov and M. V. Fedorjuk, *Semiclassical Approximation in Quantum Mechanics*, D. Reidel, Dordrecht (1981).
T. B. Maslova and A. A. Ruzmaikin, "The growth of magnetic field fluctuations in turbulent flow", *Magnitnaya gidrodinamika*, No. 3 (1987) 3–7.
T. B. Maslova, T. S. Shumkina, A. A. Ruzmaikin and D. D. Sokoloff, "Self-excitation of fluctuative magnetic fields in spatially bound random flow", *Dok. Akad. Nauk* (SSSR) **294** (1987) 1373–1376.
H. P. McKean, *Stochastic Integrals*, Academic Press (1969).
A. J. Mestel, "On the flow in a channel induction surface", *J. Fluid Mech.* **147** (1984) 431–447.
M. Meneguzzi, U. Frisch and A. Pouquet, "Helical and non-helical turbulent dynamos", *Phys. Rev. Lett.* **47** (1981) 1060–1064.
N. V. Menshikov, S. A. Molchanov and A. S. Sidorenko, "Percolation theory and its applications", *Itogi Nauki i Tekhniki: Kibernetika i Matematicheskaya Statistika*, VINITI, Moscow **24** (1986) 3–63.
A. S. Mikhailov and I. V. Uporov, "Critical effects in media with multiplication, decay and diffusion", *Usp. Fiz.*, (SSSR) **144** (1984) 79–112.
P. W. Milonni, M.-L. Shih and J. R. Ackerhalt, *Chaos in Laser-Matter Interaction*, World Scientific (1987).
H. K. Moffatt, "The degree of knottedness of tangled vortex lines", *J. Fluid Mech.* **35** (1969) 117–129.
H. K. Moffatt, *Magnetic Field Generation in Electrically Conducting Fluids*, Cambridge Univ. Press (1978).
H. K. Moffatt and M. R. E. Proctor, "Topological constraints associated with fast dynamo action", *J. Fluid Mech.* **154** (1985) 493–507.

S. A. Molchanov, L. I. Piterbarg, A. A. Ruzmaikin and D. D. Sokoloff, "Variability of temperature field of ocean surface", *Dok. Akad. Nauk* (SSSR) **283** (1985) 1801–1803.

S. A. Molchanov, A. A. Ruzmaikin and D. D. Sokoloff, "Dynamo equations in random short-correlated velocity field", *Magnitnaya gidrodinamika*, No. 4 (1983) 67–73.

S. A. Molchanov, A. A. Ruzmaikin and D. D. Sokoloff, "A Dynamo theorem", *Geophys. Astrophys. Fluid Dyn.* **30** (1984) 242–259.

S. A. Molchanov, A. A. Ruzmaikin and D. D. Sokoloff, "Kinematic dynamo in random flow", *Usp. Fiz. Nauk* (SSSR) **145** (1985) 593–628.

S. A. Molchanov, A. A. Ruzmaikin and D. D. Sokoloff, "Short-correlated random flow as the fast dynamo", *Dok. Akad. Nauk* (SSSR) **295** (1987) 576–579.

A. S. Monin and A. M. Yaglom, *Statistical Fluid Mechanics*, **2**, MIT Press (1971).

J. Nash, "C^1-isometric imbeddings", *Ann. Math.* **60** (1954) 383–396.

J. Nash, "The imbedding problem for Riemannian manifolds", *Ann. Math.* **63** (1956) 20–63.

V. V. Nemytsky and V. V. Stepanov, *Qualitative Theory of Differential Equations*, Gostekhizdat, Moscow (1947).

V. P. Nosko, "The local structure of Gaussian random fields in vicinity of high peaks", *Dok. Akad. Nauk* (SSSR) **189** (1969) 714–717.

E. A. Novikov, "Intermittency and scaling in structure of turbulent flows", *Prikl. mat. i mekh.* **35** (1971) 266–277.

V. G. Novikov, A. A. Ruzmaikin and D. D. Sokoloff, "Fast dynamo in reflexively invariant random velocity field", *JETP* **85** (1983) 909–918.

A. M. Obukhov, "Structure of temperature field in turbulent flow", *Izvestiya An SSSR*, ser. geograf. geofiz. **13** (1949) 58–96.

E. N. Parker, *Cosmical Magnetic Fields* (*Their Origin and Activity*), Clarendon Press, Oxford (1979).

Peng Zhen, *Beijing, the Capital of China*, Red Flag Publ. House, Beijing (1986).

J. Perrin , *Les Atomes*, Libraire Félix Alcan, Paris (1936).

L. S. Polak and A. S. Mikhailov, *Selforganization in Non-equilibrium Physico-chemical Systems*, Nauka, Moscow (1983).

A. M. Polyakov, "Isomeric states of quantum fields", *JETP* **68** (1975) 1975–1990.

M. M. Postnikov and A. T. Fomenko, *New Methods of Statistical Analysis of Narrative and Numerical Sources of Ancient History*, Moscow (1980).

Yu. P. Pyt'ev and I. A. Shishmaryov, *Course in Probability Theory and Mathematical Statistics for Physicists*, Moscow State Univ., Moscow (1983).

P. K. Rashevsky, "On the dogma of natural series", *Usp. Mat. Nauk* **28** (1973) 243–246.

B. D. Rozovsky, "On problem of kinematic dynamo in random flow". *IV Intern.*

Vilnius Conf. of Probability Theory and Math. Statistics, **4**, Vilnius (1985) 256–257.
D. Ruelle and F. Takens, "On the nature of turbulence". *Commun. Math. Phys.* **20** (1971) 167–192.
A. A. Ruzmaikin, A. M. Shukurov and D. D. Sokoloff, *Magnetic Fields of Galaxies*, Kluwer, Dordrecht (1988).
A. A. Ruzmaikin A. M. Shukurov and D. D. Sokoloff, *Magnetic fields of galaxies*, Kluwer Academic Publishers, Dordrecht (1988).
A. A. Ruzmaikin and D. D. Sokoloff, "Helicity, linkage and dynamo action", *Geophys. Astrophys. Fluid Dyn.* **16** (1980) 73–82.
S. M. Rytov, Ya. A. Kravtsov and V. I. Tatarskü, *Introduction to Statistical Radio Physics* **2**, Nauka, Moscow (1978).
A. D. Sakharov, "Violation of CP invariance, C-asymmetry and baryon asymmetry of the Universe", *Pisma JETP* **5** (1967) 32–35.
B. Saltzmann, "Finite amplitude free convection as an initial value problem "I", *J. Atmos. Sci.* **19** (1962) 329–341.
Yu. Saltzman, "How to seed a two-dimensional field". *Nauka i Zhizn* (Science and Life, USSR), No. 10 (1987) 103.
F. W. Schelling, *Grundiegung der Positive Philosophie*, Torino (1972).
E. Schrödinger, "Uber die Umkehrung der Naturgesetze", Sitzungsberder Preuss, *Akad. der Wiss.* (1931) 144–153.
Sextus Empiricus with an English translation by the Rev. R. G. Bury, *Litt. D.* **4**, Cambridge (1949).
S. P. Shandarin and Ya. B. Zeldovich, "Topological mapping properties of collisionless potential and vortex motion", *Phys. Rev. Lett.* **52** (1984) 1488–1491.
E. D. Siggia, "Collapse and amplification of a vortex filament", *Phys. Fluids* **28** (1985) 794–805.
J. L. Singe, *Relativity: The General Theory*, North-Holland, Amsterdam (1960).
B. M. Smirnov, "Fractal clusters", *Usp. Fiz. Nauk* (SSSR) **149** (1986) 177–219.
M. von Smoluchowski, "Zur kinetischen theorie der Brownschen molecularbewegung in der suspensionen", *Ann. der Physik* **21** (1906) 756–780.
D. D. Sokoloff and V. F. Shvartsman, "Estimate of the Universe size from topological arguments", *JETP* **66** (1974) 412–420; *Sov. Phys. JETP* **39** (1974) 196–200.
D. D. Sokoloff and M. L. Bujakaite, "On some difficulties in application of the least-squares method to the rubidium-strontium geochronology", in *Evolution of the Core-Mantle System*, Nauka, Moscow (1986) 207–216.
D. D. Sokoloff and T. S. Shumkina, "On the structure of statistical moments of a scalar field in random medium", *Vestnik Moscow Univ.*, Ser. 3, vol. **29** (1988) 23–28.
I. M. Sokolov, "Dimensions and other geometrical critical exponents in

percolation theory", *Usp. Fiz. Nauk* (SSSR) **150** (1986) 221–255.

A. M. Soward, "Fast dynamo in a steady flow", *J. Fluid Mech.* **180** (1987) 267–295.

E. M. Stuerman and M. K. Trofimova, *Slavery in the Early Roman Empire in Italy*, Nauka, Moscow (1971).

I. P. Stakhanov, *On the Physical Nature of Ball Lightning*, Energoatomizdat, Moscow (1985).

A. A. Starobinsky, "Dynamics of phase transition in the new inflationary Universe scenario and generation of perturbation", *Phys. Lett.* **B117** (1982) 175–178.

J. O. Stenflo, "Small-scale solar magnetic fields", in *Basic Mechanisms of Solar Activity*, Proc. IAU Symp. No. 17, V. Bumba and J. Kleczek (eds.), D. Reidel, Dordrecht (1976) 69–99.

R. L. Stratonovich, *Conditional Markov Processes*, Moscow State Univ. Press, Moscow (1966).

F. Takens, "Detecting strange attractors in turbulence", in *Dynamical Systems and Turbulence*, Lecture Notes in Mathematics **898** (1980) 366–381.

G. I. Taylor, "Diffusion by continuous movements", *Proc. Lond. Math. Soc.*, Ser. 2 **20** (1921) 196–211.

L. S. Tong, "Heat transfer and nuclear reactor safety", in *Proc. Sith. Jut. Heat Trans. Conf.*, Toronto **6** (1978).

S. I. Vainstein, "The dynamo of small-scale magnetic field", *JETP* **79** (1980) 2175–2183; *Sov. Phys. JETP* **52** (1980) 1099–1107.

S. I. Vainstein and Ya. B. Zeldovich, "Origin of magnetic fields in astrophysics", *Usp. Fiz. Nauk* (SSSR) **106** (1972) 431–457; *Sov. Phys. Usp.* **15** (1972) 159–172.

V. I. Vinogradov, A. M. Leites, M. I. Bujakaite and B. G. Pokrovsky, "Pb-Sr System in the rocks of Oliokmo-Kalar anorthozite massif and its northern frame", *Dok. Akad. Nauk* (SSSR) **273** (1983) 445–449.

G. A. Ville, *Biology*, W. B. Saunders (1967).

K. Weierstrass, *Abhandlungen aus der Funktionleher*, Springer, Berlin (1886).

S. B. Wesselowsky, "Gambling as a source of income in Moscovite State of 17th century", in *Selected Papers in Commemoration of W. O. Klutchewsky*, Moscow (1909) 291–316.

J. A. Wheeler, *Geometrodynamics*, Academic Press (1962).

N. Wiener, *Nonlinear Problems in Random Theory*, The Technology Press of Massachusetts Inst. of Tech. and John Wiley and Sons (1958).

Ch. Wolf, *Störfall. Nachrichten eines Tages*, Aufbau, Berlin-Weimar (1987).

J. Yorke and E. Yorke, "Metastable chaos: the transition to sustained chaotic behavior in the Lorentz model", *J. Stat. Phys.* **21** (1979) 263–278.

Ya. B. Zeldovich, "Limiting laws for free upwelling convective flows", *JETP* **7** (1937) 1463–1465.

Ya. B. Zeldovich, "The magnetic field in the two-dimensional motion of a conducting turbulent liquid", *JETP* **31** (1956) 154–156; *Sov. Phys. JETP* **4** (1957) 460–462.

Ya. B. Zeldovich, "Gravitational instability. An approximation theory for large density perturbations", *Astron. Astrophys.* **5** (1970) 84–89.

Ya. B. Zeldovich, "Exact solution of the problem of diffusion in a periodic velocity field and turbulent diffusion", *Dok. Akad. Nauk* (SSSR) **266** (1982) 821–826.

Ya. B. Zeldovich, "Percolation properties of a two-dimensional steady random magnetic field", *Pisma JETP* **38** (1983a) 51–54.

Ya. B. Zeldovich, "On disturbances leading to isolated domains", *Dok. Akad. Nauk* (SSSR) **270** (1983b) 1369–1372.

Ya. B. Zeldovich, "Topological and percolation properties of potential mapping with glueing", *Proc. Nat. Acad. Sci. USA*, Ser. Math. **80** (1983c) 2410–2411.

Ya. B. Zeldovich, G. I. Barenblatt, V. B. Librovich and G. M. Makhviladze, *Mathematical Theory of Combustion and Explosion*, Plenum (1985).

Ya. B. Zeldovich P. I. Kolykhalov, "Instability of a tangential discontinuity: the first approximation in the wave vector", *Dok. Akad. Nauk* (SSSR) **266** (1982) 302–304.

Ya. B. Zeldovich S. A. Molchanov, A. A. Ruzmaikin and D. D. Sokoloff, "Intermittency of passive fields in random media", *JETP* **89** (1985) 2061–2072.

Ya. B. Zeldovich, S. A. Molchanov, A. A. Ruzmaikin and D. D. Sokoloff, "Intermittency in random media", *Usp. Fiz.* (SSSR) **152** (1987a) 3–32.

Ya. B. Zeldovich, S. A., Molchanov, A. A. Ruzmaikin and D. D. Sokoloff, "Intermittency, diffusion and generation in a nonstationary random medium", *Sov. Sci. Rev. C. Math. Phys.* **7** (1987b) 1–107.

Ya. B. Zeldovich, S. A. Molchanov, A. A. Ruzmaikin and D. D. Sokoloff, "Self-excitation of non-linear scalar field in a random medium", *Proc. Nat. Acad. Sci. USA*, ser. Math. **84** (1987) 6323–6325.

Ya. B. Zeldovich and I. D. Novikov, *Relativistic Astrophysics*, Chicago Univ. Press (1971).

Ya. B. Zeldovich, A. A. Ruzmaikin and D. D. Sokoloff, *Magnetic Fields in Astrophysics*, Gordon and Breach (1983).

Ya. B. Zeldovich, A. A. Ruzmaikin and D. D. Sokoloff, "On representation of a three-dimensional vector field by scalar potentials", *Dok. Akad. Nauk* (SSSR) **284** (1985) 103–106.

Ya. B. Zeldovich and S. F. Shandarin, "Voids in the Universe", *Pisma Astron. J.* (SSSR) **8** (1982) 131–135.

Ya. B. Zeldovich and D. D. Sokoloff, "On zeros of eigenfunctions of free and forced oscillations in continual dissipational media", *Dok. Akad. Nauk* (SSSR) **275** (1984) 358–361.

Ya. B. Zeldovich and D. D. Sokoloff, "Fractals, similarity, intermediate asymptotics", *Usp. Fiz. Nauk* (SSSR) **146** (1985) 493–506; *Sov. Phys. Usp.* **28** (1985) 608–616.

Ya. B. Zeldovich and D. D. Sokoloff, "On eigenoscillations of a region with random boundary", *Dok. Akad. Nauk* (SSSR) **280** (1985) 75–77.

Ya. B. Zeldovich, D. D. Sokoloff and A. A. Starobinsky, "Some problems of geometry in the whole relevant to the general relativity", in *150 Years of Lobachevskian Geometry*, VINITI, Moscow (1977) 271–282.

Ya. B. Zeldovich, "Collapse of a small mass in the general relativity", *JETP* **42** (1962) 641–642.

Ya. B. Zeldovich, "Why the Universe expands?" *Priroda* (Nature USSR) No. 2 (1984) 66–71.

Ya. B. Zeldovich, "Can the Universe be formed "from nothing"?" *Priroda* (Nature USSR) No. 4 (1988) 16–26.

Ya. B. Zeldovich and A. A. Starobinsky, "A non-trivial topology Universe and possibility of its quantum birth", *Pisma Astron. J.* (SSSR) **10** (1984) 323–328.

A. Zygmund, *Trigonometrical Series*, Cambridge Univ. Press (1959).